Glimpsing Reality:
Ideas in Physics and the Link to Biology
Edited by Paul Buckley and F. David Peat

Twenty years ago Paul Buckley and F. David Peat asked several physicists, biologists, and chemists who had been involved in some of the most exciting discoveries in modern scientific thought to participate in the interviews that formed the heart of the book *A Question of Physics: Conversations in Physics and Biology*. *Glimpsing Reality* is an expanded version of that book.

The conversations – with Bohm, Pattee, Penrose, Rosen, Rosenfeld, Somorjai, Weizsäcker, Wheeler, and Nobel prizewinners Heisenberg and Dirac, cofounders of quantum theory, and Prigogine – explore issues which have shaped modern physics and ones which hint at what may form the next scientific revolution. The discussions range over a set of basic problems in physical theory and their possible solutions – the understanding of space, time, and cosmology, the genesis of quantum theory and criticism of the standard interpretations of it, quantum and relativity theories and attempts to unite them – and the conceptual links between physics and biology. The approach is nontechnical, with an emphasis on the basic assumptions of modern science and their implications for understanding the world we live in.

All of the original interviews have been preserved. An introduction has been added to expand the thematic content of the mini-introductions preceding each interview. A new conversation (between the editors) has been added, a dialogue that places the fundamental ideas of quantum theory in a broad perspective to include work on chaos theory and superstring theory. Also new to this volume are two original essays that further develop the main thrust of the text – an exploration of the boundaries between physics and biology with the supervening idea being quantum theory and problems of its interpretation.

PAUL BUCKLEY was formerly associate professor of chemistry at l'Université Laval. He has been science adviser to several government agencies and was, in 1992–3, Visiting Scientist in the Department of Physiology and Biophysics at Dalhousie University.

F. DAVID PEAT is a physicist and author of many books, including *Einstein's Moon: Bell's Theorem and the Curious Quest for Quantum Reality*; *Superstrings and the Search for the Theory of Everything*; and *Synchronicity, the Bridge between Matter and Mind*.

Glimpsing Reality:
Ideas in Physics and
the Link to Biology

EDITED BY PAUL BUCKLEY AND F. DAVID PEAT

UNIVERSITY OF TORONTO PRESS

Toronto Buffalo London

© University of Toronto Press Incorporated 1996
Toronto Buffalo London
Printed in Canada

ISBN 0-8020-0575-6 (cloth)
ISBN 0-8020-6994-0 (paper)

Revised and expanded edition of *A Question of Physics:*
Conversations in Physics and Biology
(University of Toronto Press 1979).

Printed on acid-free paper

Canadian Cataloguing in Publication Data

Main entry under title:

Glimpsing reality : ideas in physics and the link
to biology

Rev. ed.
Previously published under title: A question of
physics.
ISBN 0-8020-0575-6 (bound) ISBN 0-8020-6994-0 (pbk.)

1. Physicists – Interviews. 2. Biologists –
Interviews. I. Buckley, Paul, 1938– .
II. Peat, F. David, 1938– . III. Title: A
question of physics.

QC15.B83 1995   530   C95-932356-2

University of Toronto Press acknowledges
the financial assistance to its publishing program
of the Canada Council and the
Ontario Arts Council.

TO OUR PARENTS

Contents

Preface to the revised edition

Twenty years ago we asked several physicists, including two of the co-founders of quantum theory, Werner Heisenberg and Paul Dirac, to participate in the interviews which form the heart of this book. The present work is an expanded version of *A Question of Physics*, which appeared in 1979. All of the original interviews have been preserved without change. An introduction has been added to expand the thematic content of the mini-introductions preceding the conversations. A new conversation has been added, which is intended as a kind of update on developments since the first edition appeared, and two original essay-type contributions have been included to further develop some of the ideas presented in the main text.

Preface to the first edition

This book contains interviews with physicists, biologists, and chemists who have been involved in some of the most exciting discoveries in modern scientific thought. Some time ago we approached the Canadian Broadcasting Corporation with a proposal for a series of radio programs in which the revolutions taking place in physics during the last fifty years could be explored. The series would attempt to re-create the elation and argument, the disappointment and confusion, which physicists experienced during the origins of the quantum theory, along with some of the more exciting developments in quantum and relativity theories. By presenting science through the voices of its practitioners we hoped to convey a vivid, if at times unpolished, first-hand account. The resulting interviews are the origin of the present book, in which we have preserved the tempo and integrity of the original dialogues by indulging in the minimum amount of editing.

The success of the venture depended to a great extent upon the enthusiasm of the scientists we interviewed, and here we feel lucky in having selected physicists who have not only made important contributions to human thought but have also the ability to transmit their ideas clearly and directly.

In selecting topics for discussion we have betrayed our own prejudices. Rather than dwell upon the successes of modern physics we have explored the cracks in its fifty-year-old façade. We have concentrated on areas which, we feel, hint at the next scientific revolution. Perhaps in this context we own an apology to an important group of scientists – those engaged in elementary particle research. Some physicists feel that the search for 'ultimate building-blocks of matter' is one of the most promising modern areas of research. It was our belief, however, that there are

deeper questions to be explored, and that the goal of 'the most fundamental particle' is somewhat of a throwback to the presuppositions of
classical physics.

We have also included in this book, which is otherwise concerned with
the problems of physics, a round-table discussion on theoretical biology.
This young subject has all the intellectual challenge and excitement associated with physics in the twenties. Possibly in reading of the biologist's
responses to his present difficulties we may be better able to understand
the situation which faced physicists at a time when no atomic theory existed and there was simply an accumulation of spectroscopic data and a
new and confusing quantum principle. The discussion also provides an
example of the way in which traditional boundaries between the sciences
are erased as similar questions are raised and mathematical techniques
employed in diverse disciplines.

We hope that this book will serve as a useful overview for the practi-
tioner of science and, at the same time, give the non-scientist some understanding of the revolution which has taken place in our understanding
of the world. It was our intention to avoid technical terms and maintain a
level of discussion accessible to a broad audience, but at times the scientists we interviewed became involved in questions which have troubled
the scientific community for nearly half a century. They are to be excused
for occasionally forgetting that 'the collapse of the wave function,' 'non-
classical logic,' and 'the Copenhagen interpretation' are not topics which
the average family discusses over morning coffee. We trust that our short
appendix will be helpful in providing a background for such questions.

For assistance in the preparation of the manuscript, we extend our
deepest thanks to Jane Wykes, who cheerfully undertook the arduous
task of verifying the transcripts with the original tapes and typed the first
edited version.

PAUL BUCKLEY
F. DAVID PEAT

Introduction to the revised edition

One of the things which makes physical theory intellectually attractive to many persons is the great amount of discussion which characterized quantum theory during its early stages and which continues enthusiastically, if less intensely, at the present time. Today there is fresh discussion centred on the foundations of quantum mechanics, and it appears that those foundations are not as firm as one had earlier thought. Though quantum mechanics is, in its formalism and in its detailed practice, extremely hard edged and very successful, it does invite alternative interpretations which are competing for attention. It seems to some that the revolution in science, which accompanied the early days of the century that is drawing to a close, is not yet finished despite the outstanding efforts of many great minds. But one really need not ask whether the revolution is finished for there are always the voices of dissent, and good questions keep getting asked even if decades go by before true attention is paid to them. All this just adds to the excitement which currently prevails in physics and which likely influences the work of other disciplines and activities. At the same time, however, one finds a note of seriousness, not to say unease, as a characteristic of the present mood.

Werner Heisenberg and Leon Rosenfeld are this book's spokespersons for the standard interpretation, also known as the Copenhagen interpretation, and they comment extensively upon its characteristics. Heisenberg calls the interpretation abstract, and possibly this has been a stumbling-block for some physicists, though the limitations of using ordinary language in physical descriptions are evident. Heisenberg and Rosenfeld communicate very clearly their sensitivities on the issue of language and the boundaries of classical concepts. They also give us many insights on the conditions of the origin of the quantum theory, thus leaving us with valuable pointers for contemporary studies in the history and philosophy of science. Of course, the interpretation may be formulated more rigorously than can be set out in these two interviews, but most would admit that it is good to hear the story from those near the centre of the action in the discussions of the 1920s in Copenhagen, Göttingen, and other European cities.

There is one alternative interpretation which has been recently pointed to in the pages of *Scientific American* (May 1994) and this is David Bohm's 'ontological interpretation.' It is a serious contender among some theoretical physicists, biding its time until the day arrives and the orthodox interpretation no longer maintains its general acceptance. In his interview David Bohm explores meaning in a probing critique of attitudes in physics. Once again the question of the use of language is brought into the foreground where it belongs. The feeling that there is something newly positive about quantum theory is exemplified by Carl Weizsäcker, who directs our attention to tense logic and to the issue of human time. He also discusses historical perspectives as they might apply to our period and this theory.

Paul Dirac surely represents the many physicists who remain untroubled by problems of interpretation; in his interview he resolutely refuses to talk about it. He prefers instead to comment upon certain cosmological issues, but he does make a few remarks on the then current state of theoretical physics. We are especially glad to have this interview as it is one of the very few that he consented to give. The interview with Roger Penrose introduces some of his imaginative work on twistors and the nature of space-time. John Wheeler ranges over geometrical ideas of space and time and also opens up some of the inner feelings possible in science, pointing toward its beauty. Wheeler also believes that the quantum theory allows us to feel that we are participating with Nature in the unfolding universe.

Ilya Prigogine firmly states a belief in participation based upon his results in the field of irreversible thermodynamics involving dissipative structures. This work seems to open out toward a biological frame and the more sophisticated notions of order which life sustains. Life itself is the subject of the mini-symposium involving Howard Pattee, Robert Rosen, Raymond Somorjai, and the co-authors. It becomes the locus of a spirited search for coherence in life's complexity and how best to grasp it. They attempt to demonstrate how physics and biology might relate in a more fruitful way than is found at present.

Along with its sharp edges and hard-won understanding, quantum mechanics may stimulate new feelings of participation in Nature. 'Evolution and Quantum Consciousness' by Paul Buckley is a series of reflections studying the implications of the coexistence of a theory of evolution and a quantum theory. In his essay, 'The Schrödinger Question: What Is Life? Fifty Years Later,' Robert Rosen examines some of the consequences for biology of asking this question today in Schrödinger's own manner. And because Erwin Schrödinger is one of the co-founders of the quantum theory this makes another connection for us between the sciences of matter and the sciences of life which are presented in this book.

Conversations

Werner Heisenberg

While a student of Arnold Sommerfeld at Munich in the early 1920s Werner Heisenberg (1901–75) first met the Danish physicist Niels Bohr. He and Bohr went for long hikes in the mountains and discussed the failure of existing theories to account for the new experimental results on the quantum structure of matter. Following these discussions Heisenberg plunged into several months of intensive theoretical research but met with continual frustration. Finally, suffering from a severe attack of hay fever, he retreated to the treeless island of Helgoland. After days spent relaxing and swimming Heisenberg suddenly experienced the giddy sensation of looking down into the heart of nature and conceived the basis of the quantum theory. He took this theory to Bohr at Copenhagen, and for the next few weeks they argued and probed its implications long into the night. The results of these discussions became known as the 'Copenhagen interpretation of quantum theory' and are accepted by most physicists. Aspects of the interpretation include Heisenberg's uncertainty principle and Bohr's principle of complementarity.

Heisenberg made other important discoveries in physics, and became one of the most distinguished physicists of the century. He was awarded the Nobel Prize for Physics in 1932. His scientific attitudes reflect a debt to philosophy and in particular his respect for Plato. Some of his thoughts on science and society are recorded in a readable autobiography entitled *Physics and Beyond.*

In recent years Heisenberg adopted the unpopular position of criticizing research in elementary particle physics and proposing that symmetries and not elementary particles form the fundamental starting-point for a description of the world. Towards the end of this chapter he touches upon this theory and its reception.

Professor Heisenberg was interviewed one sunny morning in his office at the Max Planck Institute in Munich. We began by asking Heisenberg to recall the early days of quantum theory but it became apparent that great men have no desire to live in the past and he was just as eager to talk about the future of physics.

DP *Could you reminisce about the time when you arrived at the idea of quantum mechanics?*

At that time, there was general discussion among young physicists about the possible ways to establish a coherent quantum theory, a coherent quantum mechanics. Among the many attempts, the most interesting for me was the attempt of H.A. Kramers to study the dispersion of atoms and, by doing so, to get some information about the amplitudes for the radiation of atoms. In this connection, it occurred to me that in the mathematical scheme these amplitudes behaved like the elements of a mathematical quantity called a matrix. So I tried to apply a mathematical calculus to the experiments of Kramers, and the more general mechanical models of the atom, which later turned out to be matrix mechanics. It so happened at that time I became a bit ill and had to spend a holiday on an island to be free from hay fever. It was there, having good time to think over the questions, that I really came to this scheme of quantum mechanics and tried to develop it in a closed mathematical form.

My first step was to take it to W. Pauli, a good friend of mine, and to discuss it with him, then to Max Born in Göttingen. Actually, Max Born and Pascual Jordan succeeded in giving a much better shape and more elegant form to the mathematical scheme. From the mathematical relations I had written down, they derived the so-called commutation relations. So, through the work of Born and Jordan, and later Paul Dirac, the whole thing developed very quickly into a closed mathematical scheme.

I also went to discuss it with Niels Bohr, but I can't be sure whether this was in July, August, or September of that year [1925].

Half a year later the first papers of E. Schrödinger became known. Schrödinger tried to develop an older idea of Louis de Broglie into a new mathematical scheme, which he called wave mechanics. He was actually able to treat the hydrogen atom on the basis of his wave mechanical scheme and, in the summer of 1926, he was also able to demonstrate that his mathematical scheme and matrix mechanics were actually two equivalent mathematical schemes, that they could be simply translated into each other. After that time, we all felt that this must be the final mathematical form of quantum theory.

DP *Had you and Bohr begun the interpretation of this work before Schrödinger's paper came out?*

Of course, there was continuous discussion, but only after Schrödinger's paper did we have a new basis for discussion, a new basis for interpreting quantum theory. In the beginning there was strong disagreement between Schrödinger and ourselves, *not* about the mathematical scheme, but about its interpretation in physical terms. Schrödinger thought that by his work physics could again resume a shape which could well be compared with Maxwell's theory or Newton's mechanics, whereas we felt that this was not possible. Through long discussions between Bohr and Schrödinger in the fall of 1926, it became apparent that Schrödinger's hopes could not be fulfilled, that one needed a new interpretation. Finally, from these discussions, we came to the idea of the uncertainty relations, and the rather abstract interpretation of the theory.

PB *Did Schrödinger ever like that interpretation?*

He always disliked it. I would even guess that he was not convinced. He probably thought that the interpretation which Bohr and I had found in Copenhagen was correct in so far as it would always give the correct results in experiments; still he didn't like the language we used in connection with the interpretation. Besides Schrödinger, there were also Einstein, M. von Laue, M. Planck, and others who did not like this kind of interpretation. They felt it was too abstract, and too far removed from the older ideas of physics. But, as you know, this interpretation has, at least so far, stood the test of all experiments, whether people like it or not.

PB *Einstein never really liked it, even until the day he died, did he?*

I saw Einstein in Princeton a few months before his death. We discussed quantum theory through one whole afternoon, but we could not agree on the interpretation. He agreed about the experimental tests of quantum mechanics, but he disliked the interpretation.

DP *I felt that at some point there was a slight divergence between your views and Bohr's, although together you are credited with the Copenhagen interpretation of quantum mechanics.*

That is quite true, but the divergence concerned more the method by which the interpretation was found than the interpretation itself. My point of view was that, from the mathematical scheme of quantum mechanics, we had at least a partial interpretation, inasmuch as we can say, for instance, that those eigenvalues which we determine *are* the energy values of the discrete stationary states, or those amplitudes which we determine are responsible for the intensities of the emitted lines, and so on. I believed it must be possible, by just extending this partial interpretation, to get to a complete interpretation. Following this way of thinking, I came to the uncertainty relations.

Now, Bohr had taken a different starting-point. He had started with the dualism between waves and particles – the waves of Schrödinger and the particles in quantum mechanics – and tried, from this dualism, to introduce the term *complementarity*, which was sufficiently abstract to meet the situation. At first we both felt there was a real discrepancy between the two interpretations, but later we saw that they were identical. For three or four weeks there was a real difference of opinion between Bohr and myself, but that turned out to be irrelevant.

DP *Did this have its origin in your different philosophical approaches?*

That may be. Bohr's mind was formed by pragmatism to some extent, I would say. He had lived in England for a longer period and discussed things with British physicists, so he had a pragmatic attitude which all the Anglo-Saxon physicists had. My mind was formed by studying philosophy, Plato and that sort of thing. This gives a different attitude. Bohr was perhaps somewhat surprised that one should finally have a very simple mathematical scheme which could cover the whole field of quantum theory. He would probably have expected that one would never get such a self-consistent mathematical scheme, that one would always be bound to use different concepts for different experiments, and that physics would always remain in that somewhat vague state in which it was at the beginning of the 1920s.

DP *In the interpretation you gave at that time, you seemed to imply that there did exist an ideal path and that somehow the act of measuring disturbed the path. This is not quite the same as the interpretation that you hold now, is it?*

I will say that for us, that is for Bohr and myself, the most important step was to see that our language is not sufficient to describe the situation. A word such as *path* is quite understandable in the ordinary realm of physics when we are dealing with stones, or grass, etc., but it is not really understandable when it has to do with electrons. In a cloud chamber, for instance, what we see is *not* the path of an electron, but, if we are quite honest, only a sequence of water droplets in the chamber. Of course we like to interpret this sequence as a path of the electron, but this interpretation is only possible with restricted use of such words as *position* and *velocity*. So the decisive step was to see that all those words we used in classical physics – *position*, *velocity*, *energy*, *temperature*, etc. – have only a limited range of applicability.

The point is we are bound up with a language, we are hanging in the language. If we want to do physics, we must describe our experiments and the results to other physicists, so that they can be verified or checked by others. At the same time, we know that the words we use to describe the experiments have only a limited range of applicability. That is a fundamental paradox which we have to confront. We cannot avoid it; we have simply to cope with it, to find what is the best thing we can do about it.

DP *Would you go so far as to say that the language has actually set a limit to our domain of understanding in quantum mechanics?*

I would say that the concepts of classical physics which we necessarily must use to describe our experiments do not apply to the smallest particles, the electrons or the atoms – at least not accurately. They apply perhaps qualitatively, but we do not know what we mean by these words.

Niels Bohr liked to tell the story about the small boy who comes into a shop with two pennies in his hands and asks the shopkeeper for some mixed sweets for the two pennies. The shopkeeper gives him two sweets and says 'You can do the mixing yourself.' This story, of course, is just meant to explain that the word *mixing* loses its meaning when we have only two objects. In the same sense, such words as *position* and *velocity* and *temperature* lose their meaning when we get down to the smallest particles.

DP *The philosopher Ludwig Wittgenstein originally started off by thinking that words were related to facts in the world, then later reversed his position to conclude that the meaning of words lay in their use. Is this reflected in quantum mechanics?*

I should first state my own opinion about Wittgenstein's philosophy. I never could do too much with early Wittgenstein and the philosophy of the *Tractatus Logico-philosophicus*, but I like very much the later ideas of Wittgenstein and his philosophy about language. In the *Tractatus*, which I thought too narrow, he always thought that words have a well-defined meaning, but I think that is an illusion. Words have no well-defined meaning. We can sometimes by axioms give a precise meaning to words, but still we never know how these precise words correspond to reality, whether they fit reality or not. We cannot help the fundamental situation – that words are meant as a connection between reality and ourselves – but we can never know how well these words or concepts fit reality. This can be seen in Wittgenstein's later work. I always found it strange, when discussing such matters with Bertrand Russell, that he held the opposite view; he liked the early work of Wittgenstein and could do nothing whatsoever with the late work. On these matters we always disagreed, Russell and I.

I would say that Wittgenstein, in view of his later works, would have realized that when we use such words as *position* or *velocity*, for atoms, for example, we cannot know how far these terms take us, to what extent they are applicable. By using these words, we learn their limitations.

DP *Would it be true to say that quantum mechanics has modified language, and, in turn, language will re-modify the interpretation of quantum mechanics?*

There I would not quite agree. In the case of relativity theory, I would agree that physicists have simply modified their language; for instance, they would use the word *simultaneous* now with respect to certain coordinate systems. In this way they can adapt their language to the mathematical scheme. But in quantum theory this has not happened. Physicists have never really tried to adapt their language, though there have been some theoretical attempts. But it was found that if we wanted to adapt the language to the quantum theoretical mathematical scheme, we would have to change even our Aristotelian logic. That is so disagreeable that nobody wants to do it; it is better to use the words in their limited senses, and when we must go into the details, we just withdraw into the mathematical scheme.

I would hope that philosophers and all scientists will learn from this change which has occurred in quantum theory. We have learned that language is a dangerous instrument to use, and this fact will certainly have its repercussions in other fields, but this is a very long process which will last through many decades I should say.

Even in the old times philosophers realized that language is limited; they have always been sceptical about the unlimited use of language. However, these doubts or difficulties have, perhaps, been enhanced through the present developments in physics. I might mention that most biologists today still use the language and the way of thinking of classical mechanics; that is, they describe their molecules as if the parts of the molecules were just stones or something like that. They have not taken notice of the changes which have occurred in quantum theory. So far as they get along with it, there is nothing to say against it, but I feel that sooner or later, also in biology, one will come to realize that this simple use of pictures, models, and so on will not be quite correct.

PB *At what point does the transition occur from the non-path to the path in a biological system? Is a DNA molecule already a classical object, or is a cell a classical object?*

There is, of course, not a very well defined boundary; it is a continuous change. When we get to these very small dimensions we must be prepared for limitations. I could not suggest any well-defined point where I have to give up the use of a word. It's like the word *mixing* in the story; you cannot say 'when I have two things, then I can mix them.' But what if you have five or ten? Can you mix then?

PB *It seems to me that there is something very important here about language. We are living beings formed from coherent structures like DNA and we apparently have classical paths and our existence is understandable within this language. But then we can analyse by reducing these complex, coherent wholes to smaller and smaller parts, and is it not perhaps this process of reduction that is at the root of the paradox?*

I would say that the root of the difficulty is the fact that our language is formed from our continuous exchange with the outer world. We are a part of this world, and that we have a language is a primary fact of our life. This language is made so that in daily life we get along with the world; it cannot be made so that, in such extreme situations as atomic physics, or distant stars, it is equally suited. This would be asking too much.

PB *Is there a fundamental level of reality?*

That is just the point; I do not know what the words *fundamental reality* mean. They are taken from our daily life situation where they have a good meaning, but when we use such terms we are usually extrapolating from our daily lives into an area very remote from it, where we cannot expect the words to have a meaning. This is perhaps one of the fundamental difficulties of philosophy: that our thinking hangs in the language. Anyway, we are forced to use the words so far as we can; we try to extend their use to the utmost, and then we get into situations in which they have no meaning.

DP *In discussing the 'collapse of the wave function' you introduced the notion of potentiality. Would you elaborate on this idea?*

The question is: 'What does a wave function actually describe?' In old physics, the mathematical scheme described a system as it was, there in space and time. One could call this an objective description of the system. But in quantum theory the wave function cannot be called a description of an objective system, but rather a description of observational situations. When we have a wave function, we cannot yet know what will happen in an experiment; we must also know the experimental arrangement. When we have the wave function *and* the experimental arrangement for the special case considered, only then can we make predictions. So, in that sense, I like to call the wave function a description of the potentialities of the system.

DP *Then the interaction with the apparatus would be a potentiality coming into actuality?*

Yes.

DP *May I ask you about the Kantian notion of the 'a priori,' an idea which you introduced, in a modified sense, into your discussions of quantum theory.*

As I understand the idea of 'a priori,' it stresses the point that our knowledge is not simply empirical, that is, derived from information obtained from the outer world through the senses and changed into data in the content of our brain. Rather, 'a priori' means that experience is only possible when we already have some concepts which are the precondition of experience. Without these concepts (for instance, the concepts of space and time in Kant's philosophy), we would not even be able to speak about experience.

Kant made the point that our experience has two sources: one source is the outer world (that is, the information received by the senses), and the other is the existence of concepts by which we can talk about these experiences. This idea is also borne out in quantum theory.

PB *But these concepts are part of the world also.*

Whether they belong to the world, that is hard to say; we can say that they belong to our way of dealing with the world.

PB *But we belong to the world, so, in a sense, these activities of ours also belong to the world.*

In that sense, yes.

DP *You modified the 'a priori' by introducing it as a limited concept, is that true?*

Of course, Kant would have taken the 'a priori' as something more absolute than we would do in quantum theory. For instance, Kant would perhaps have said that Euclidean geometry would be a necessary basis for describing the world, while *we*, after relativity, would say that we need not necessarily use Euclidean geometry; we can use Riemannian geometry, etc. In the same way, causality was taken by Kant as a condition for science. He says that if we cannot conclude from some fact that something must have been before this fact, then we do not know anything, and we cannot make observations, because every observation supposes that there is a causal chain connecting that which we immediately experience to that which has happened. If this causal chain does not exist, then we do not know what we have observed, says Kant. Quantum theory does not agree with this idea, and in fact proves that we can even work in cases where this causal chain does not exist.

DP *In a recent theory of yours, is not causality retained, perhaps in a new form?*

We have causality in that sense – that in order to influence something, there must be an action from one point to the next point; no action can happen if there is not this connection. But at this point one gets into rather complicated details.

DP *But, even so, you do have causality predicated on the idea of separation and action, so this again comes back to a philosophical level: what you mean by separation, and by interaction.*

We must speak about 'interaction' and 'separation,' that is quite true, and we use the terms as we did in classical theory. But, again, we see limitation. Complete separation of two events may be possible in classical theory; it is not possible in quantum theory. So we use the terms together with the fact of their limitation.

DP *What exactly are the criteria for something to be classical?*

I would say the criteria are simply that we can get along with these concepts (e.g. 'position,' 'velocity,' 'temperature,' 'energy'), and so long as we get along with them, then we are in the classical domain. But when the concepts are not sufficient, then we must say that we have gone beyond this classical domain.

Every system in physics (forget for the moment about biological systems) is always quantum theoretical, in the sense that we believe that quantum theory gives the correct answers for its behaviour. When we say that it is classical, we mean that we do get the correct or the necessary answers by using classical concepts (at least in that approximation in which we can describe the system by classical concepts). So a system is classical only within certain limits and these limits can be defined.

DP *How would you include things like irreversibility?*

Thermodynamics is a field which goes beyond Newtonian mechanics, inasmuch as it introduces the idea of thermodynamic equilibrium, or canonical distribution as W. Gibbs has put it. Thermodynamics leaves classical physics and goes into the region of quantum theory, for it speaks about situations of observation; it does not speak about the system as it is, but about the system in a certain state of being observed, namely in the state of temperature equilibrium. If this equilibrium is not obeyed, then we cannot use thermodynamics. So the whole concept of irreversibility is bound up with the concept of thermodynamic equilibrium.

DP *And is this ultimately connected with the idea of a classical limit to something? I am thinking of the measurement problem that always seems to be associated with an irreversible process: that we have a definite result for a quantum mechanical system where the quantum mechanics itself doesn't seem to predict a definite result. That is, the idea of a quantum mechanical measurement seems to be tied up with the idea of an irreversible trend.*

Yes, to some extent, because on the side of the observer we do use classical concepts. The idea that we do observe something already indicates something irreversible. If we draw a pencil line on a paper, for instance, we have established something which cannot be undone, so to speak. Every observation is irreversible, because we have gained information that cannot be forgotten.

DP *To what extent is this related to the symmetry-breaking of the quantum mechanical system where one gets classical observables?*

I would not like to connect it with symmetry-breaking; that is going a bit far. We try to describe the observational situation by writing down a wave function for the object and the equipment which is in interaction with this wave function. Just by using classical words for the equipment, we have already made the assumption of irreversibility. Or we make the assumption of statistical behaviour, because the mere use of classical words for this observation on the side of the system makes it impossible to know the total wave function of object and equipment. But we cannot use quantum theory for the equipment in a strict sense, because if we wrote down the wave function for the object *and* the equipment, we could not use classical words for the equipment, so we would not observe anything. We *do* observe only when we use classical concepts, and just at this point this hypothesis of disorder, of statistical behaviour, comes in.

DP *With regard to something like ferromagnetism, the quantum mechanical system has given rise to a macroscopic ordering. Is it true to say that a quantum mechanical system has actually broken its own symmetry and given rise to a classical variable, without any talk about a measuring apparatus, or anything exterior to the system?*

Let us consider a ferromagnet as isolated from the rest of the world for some time, and then ask what the lowest state of the system is. We find, from the quantum mechanical calculations, that the lowest state is one in which the whole system has a very large component of magnetic momentum. If we then ask 'what do we observe when we consider this system?' we see that it is convenient to ascribe the classical variable 'magnetic momentum' to the system. So we can use classical terms to describe this quantum mechanical behaviour. But this is not really a problem of observation, only a problem of how the lowest state of the system is defined.

PB *How does quantum mechanics deal with time flow or does it in fact say anything at all about it?*

I would have to repeat what C. von Weizsäcker said in his papers: that time is the precondition of quantum mechanics, because we want to go from one experiment to another, that is from one time to another. But this is too complicated to go into in detail. I would simply say that the concept of time is really a precondition of quantum theory.

PB *In the domain where quantum mechanics operates, all of the equations are reversible with respect to time, except for one experiment I believe. So time has more to do with macroscopic classical systems than microscopic quantum systems.*

I would say that irreversibility of time has to do with this other system, with those problems which I. Prigogine describes in his papers, and is certainly extremely important for the macroscopic application of quantum theory, and also for biology, of course.

DP *Can we talk about a new theory of yours, the non-linear theory of elementary particles? Are you ultimately going to introduce things like gravitation into this theory, and go over to a picture in which space and time emerge?*

Again, we have a similar situation as in ferromagnetism. We try to solve the quantum mechanical, or quantum theoretical equation, but we can see that the system acquires properties which then can be described by classical language (e.g. like speaking of a magnetic momentum, etc.). We are hoping that such phenomena as electromagnetic radiation and gravitation also can come out of the theory of elementary particles, and we have reasons to believe that this is so.

DP *The idea of symmetry is a very important part of your theory.*

Let's begin more simply by speaking about quantum mechanics, disregarding now the difficulties of elementary particle physics. In quantum mechanics we see that macroscopic bodies have very complicated properties, complicated shapes and chemical behaviour and so on. Coming down to smaller and smaller particles, we finally come to objects which are really very much simpler, for example the stationary states of a hydrogen atom. We describe its properties by saying that these states are a representation of the fundamental symmetries, such as rotation in space. So when we describe a system by writing down a few quantum numbers (in hydrogen atoms, we have the principal quantum number and the angular momentum number) this means that we know nothing except to say that this object is a representation of symmetries. The quantum numbers tell us which kind of symmetries we mean; the numbers themselves say that this object has these special properties. Thus, when we come to the smallest objects in the world, we characterize them in quantum mechanics just by their symmetry, or as representations of symmetries, and not by specifying properties such as shape or size.

DP *There are symmetries that are not related to operations in the world, e.g. the internal symmetries such as isospin. What meaning do they have? Do you think they are related ultimately to the properties of space and time?*

I suspect that isospin is a symmetry similar to space and time. I cannot say that it is related to them. I would say that there are a number of fundamental symmetries in this world which may in future be reduced to something still simpler, but so far we must take them as given, as a result of our experiments. One of the most fundamental symmetries is the symmetry of the Lorentz group, that is space and time, and then isospin groups, scale groups, and so on. So there are a number of groups which are fundamental in the sense that in describing the smallest particles we refer to their behaviour and transformations.

The idea is that one can distinguish between a natural law, a fundamental law, which determines for instance a spectrum of elementary particles, and the general behaviour of the cosmos, which is perhaps something not at once given through this law. I might remind you, for instance, of Einstein's equations of gravitation. Einstein wrote down his field equations and thought that gravitational fields are always determined by them. But the cosmos is not unambiguously determined by these field equations, although there are several models of the cosmos which are compatible with them. In the same sense, I would say that there is an underlying natural law which determines the spectrum of elementary particles, but the shape of the cosmos is not unambiguously determined by this law. Logically, it would be possible to have various types of cosmos which are in agreement with it. However, if a certain cosmological model has been 'chosen,' then this model, of course, has some consequences for the spectrum of elementary particles.

DP *Are you saying that there exist laws which are independent or outside the universe, outside the world, which reality breaks, or that it breaks the symmetry represented by the laws?*

'Laws' just means that some fundamental symmetries are inherent either in nature or in our observation of nature. You may know about the attempts of Weizsäcker, who tried to derive the laws simply from logic. We have to use language to arrive at conclusions, to study alternatives, and he questions whether from the alternatives alone we can arrive at these symmetries. I don't know whether his attempts are successful or not. In physics, we can only work with the assumption that we have natural laws. If we have no natural laws, then anything can happen, and we can only describe what we see, and that's all.

DP *Another feature of your theory which seems to go against the current trend – partons and quarks, etc. – is that you feel that no particle is any more elementary than any other.*

Even if quarks should be found (and I do not believe that they will be), they will not be more elementary than other particles, since a quark could be considered as consisting of two quarks and one anti-quark, and so on. I think we have learned from experiments that by getting to smaller and smaller units, we do not come to fundamental units, or indivisible units, but we *do* come to a point where division has no meaning. This is a result of the experiments of the last twenty years, and I am afraid that some physicists simply ignore this experimental fact.

DP *So it would seem that elementary particles are just representations of symmetries. Would you say that they are not fundamental things in themselves, or 'building-blocks of the universe,' to use the old-fashioned language?*

Again, the difficulty is in the meaning of the words. Words like *building-blocks* or *really existing* are too indefinite in their meaning, so I would hesitate to answer your questions, since an answer would depend on the definitions of the words.

DP *To be more precise, ultimately could one have a description of nature which needed* only *elementary particles or, alternatively, a description in which the elementary particles would be defined in terms of the rest of the universe? Or is there no starting-point, as it were, no single axiom on which one can build the whole of physics?*

No. Even if, for instance, that formula which Pauli and I wrote down fifty years ago turned out to be the correct formulation for the spectrum of elementary particles, it is certainly not the basis for all of physics. Physics can never be closed, or brought to an end, so that we must turn to biology or such things. What we can hope for, I think, is that we may get an explanation of the spectrum of elementary particles, and with it also an explanation of electromagnetism and gravitation, in the same sense as we get an explanation of the spectrum of a big molecule from the Schrödinger equation.

This does not mean that thereby physics has come to an end. It means that, for instance, at the boundary between physics and biology, there may be new features coming in which are not thought of in physics and chemistry. Something entirely new must happen when I try to use quantum theory within the realm of biology. Therefore I criticize those formulations which imply an end to physics.

DP *Is it ever possible to reduce physics or any element of physics purely to logic and axioms?*

I would say that certain parts of physics can always be reduced to logical mathematics or mathematical schemes. This has been possible for Newtonian physics, for quantum mechanics, and so on, so I do not doubt that it will also be possible for the world of the elementary particles. In astrophysics today, one comes upon pulsars and black holes, two regions in which gravitation becomes enormous, and perhaps a stronger force than all other forces. I could well imagine that in such black holes, for instance (if they exist), the spectrum of elementary particles would be quite different from the spectrum we now have. In the black holes, then, we would have a new area of physics, to some extent separated from that part which we now call elementary particle physics. There would be connections, and one would have to study how to go from the one to the other; but I do not believe in an end of physics.

Leon Rosenfeld

The Copenhagen interpretation of quantum theory, which grew out of discussions between Niels Bohr and Werner Heisenberg, included Leon Rosenfeld (1904–75) as one of its major proponents. Born in Charleroi, Belgium, Rosenfeld made his intellectual home in the Copenhagen of Bohr. In addition to his discoveries in theoretical physics Rosenfeld became the major apologist of the Copenhagen school after Bohr's death.

Our interview with Professor Rosenfeld took place in Copenhagen and was one of his first activities after suffering a heart attack. Later we dined with the Rosenfelds and, after spirited discussions, toasted the memory of his friend and colleague Niels Bohr.

Rosenfeld's contribution to this book is important since it deals with the interpretation of quantum theory and is possibly his last exposition on this topic. In entering a world in which the properties of an object appear to change as a result of their observation, scientists were forced to abandon their comfortable belief in material 'entities' which 'possessed' particular properties. The Copenhagen interpretation is an attempt to give an account of this new world which is intellectually satisfying and avoids the pitfalls and paradoxes generated by earlier attempts to understand the quantum theory. Professor Rosenfeld discusses the way in which we must treat knowledge of the world below the atom.

From a historical point of view Rosenfeld makes interesting observations when he discusses the differences in approach between Bohr and Heisenberg, which may have led to subtle divergences in their interpretation of quantum theory. These differences are hotly denied by Heisenberg in his interview.

PB *Professor Rosenfeld, you worked closely with Niels Bohr for many years. Could we begin with some personal reminiscences of him?*

When I first knew Niels Bohr in the 1930s there were not so many of us working in the institute, perhaps only half a dozen. He would come up every morning, since his house was near the institute, and if he met us on the stairs by any chance, the conversation could continue on the stairs for hours, or indeed at any place in the institute.

We learned that it was by those conversations that he could express himself. Whenever he had to write something down, being so anxious about complementarity, he felt that the statement contained in the first part of the sentence had to be corrected by an opposite statement at the end of the sentence. That made writing a paper a terrible business. But in conversation, it was easier; we could interrupt him, and put questions to him.

He would become completely lost in thought, even to the point of not realizing where he was. He took a walk with Klein on the day of Klein's wedding, and they nearly arrived too late! I remember I frequently travelled between Belgium, where I had my job, and Copenhagen, and once I had put my wife and child on the train, and I had the tickets in my pocket, when Bohr called to tell me an idea which he just had some hours before and which he wanted me to know before I departed. I was torn away from this conversation by the call of the station-master announcing the departure of the train. I still remember my wife's face!

PB *What was your first work with Bohr?*

The first thing I did was to help him to write down his Faraday lecture. You see, his method was to dictate a sentence, as an experiment, and then this sentence was contemplated and criticized and changed and fussed over, and so on. I was supposed to react to each sentence, to criticize, etc. That was one kind of work I did for him.

Then, very soon afterwards, there came a paper by L. Landau and R. Peierls which raised a very fundamental question about whether the field concept could be given a consistent meaning in quantum theory. In our work on this problem at first nothing could be written down, because we knew nothing at all in the beginning. We did not know whether the answer would be yes or no, and, in fact, we did not know that before the thirteenth or fourteenth proof. Every word was weighed, and every sentence has a subtle meaning, subtle in the sense that complementarity is always underlying the whole thing.

DP *People speak about the Copenhagen interpretation with very different meanings. Could you outline what Bohr really meant?*

The phrase 'Copenhagen interpretation' is actually a misnomer, in the sense that there is only one interpretation of quantum mechanics. Bohr would rather say that quantum mechanics is a whole. It is a formalism of course, a mathematical formalism, but it is also a physical theory, and therefore definite physical meanings are attached to the symbols. It is only when you take the whole thing, that is, the formalism and the meanings attached to the symbols, that you have a physical theory. The misunderstandings that have been expressed so vociferously from various sides are based on a disregard of this circumstance. They take the formalism and then they try to put upon it what they call an interpretation, without reflecting that the way in which the equations are written already implies a definite interpretation, that is, a definite relationship between the symbols and physical concepts. It is not arbitrarily that Heisenberg constructed those matrix equations and those commutation rules. He was forced to those commutation rules. In fact, he did not know beforehand that such a non-commutative algebra would come out of his effort to give a mathematical form to a clear physical idea, that is, the idea of correspondence. Therefore, you can approach this conceptual aspect of quantum mechanics only historically, because it is very much conditioned by the way in which the pioneers (Heisenberg foremost, and then Dirac very soon afterwards, and Bohr) were led to this curious new use of mathematics in order to translate into mathematical language physical ideas which were clearer than the mathematics, and not the other way around.

You see, when you first approach quantum mechanics, as a student, it is reasonable that your first effort is to understand the equations and how to handle them. And then you ask: what is the meaning of all this? And if you are for some reason afraid of statistics or of probability, then you ask yourself: could it perhaps be otherwise? That was D. Bohm's way, actually. He gave a lecture on quantum mechanics (probably the first one that he gave on this subject) and he made a book out of it. This is a very good book, a very good exposition of quantum mechanics. But it was in the process of writing the book that he had doubts about the whole thing. However, his attitude was such that he put mathematics first and he tried to hang the physics onto the mathematics, without thinking that the natural process was just the opposite.

DP *I don't think it would be fair to say that this has been Bohm's view for the last ten years.*

No. Bohm, because he is a very serious and honest thinker (and I respect him very much), at long last realized that his first approach simply did not work. But that did not mean that he was converted, like Paul on the road to Damascus. He had no such stroke as Paul had on that road. He kept his original attitude of mind, and he is now trying something different, but still always with the same outlook, which I call an idealistic outlook. He gives primacy to building concepts out of nothing. I mean, the mind is able to build any constellation of concepts.

PB *But it is ultimately based on experience too.*

Of course, I quite agree. But those I call idealists do not go so far. They stop there; they stop at the concepts, and they say that these are the primary building-stones. This is an illusion. But that's Bohm's outlook and he is now trying to dig deeper into the analysis of the concept of space and time, based on altering events. This is certainly a very true analysis; at least it has a great element of truth, I think. Nevertheless, I believe it's not the way that has proved successful in getting to a new insight in physics.

DP *Heisenberg and Bohr theories are spoken of together by many people as the Copenhagen interpretation. But I feel there is a difference.*

Heisenberg had been trained in the German school by A. Sommerfeld, not only by Sommerfeld but by the whole German environment of the time, in which great weight was put on a philosophy – the Kantian philosophy – which happened to be the dominating philosophy, whereas Bohr was quite immune, that is, he had not been exposed to Kantian philosophy. However, he, also, had followed a course of philosophy at the university, which was given by Høffding, a Danish philosopher. I would not call Høffding an eclectic, but rather he looked upon philosophy as being *au-dessus de la mêlée*, that is, outside the province of philosophers. In fact, in the preface to his course he said: 'I am concerned in this course to present the philosophical *problems*, and not the solutions, because the solutions come and go, but the problems remain.' Bohr was therefore protected from any dogmatism or any reliance upon a priori ideas. He insisted upon first understanding the physics, and then trying to put it into a mathematical form.

DP *Did Bohr not have great respect for William James's psychology?*

Yes, but I think that Bohr knew very little – practically nothing – about William James until 1935 or 1936. I remember at that time he was great friends with one of his colleagues, Rubin, one of the Gestalt psychologists. In a conversation, Rubin said to him: 'what you tell me reminds me very much of William James.' It's simply a coincidence of attitudes. James came first of course. James had very much the same approach to psychology as Bohr had to physics; that's all one can say. So when Bohr got the copy of James's treatise from Rubin, and read the chapter on the stream of thought, he was quite enthusiastic about it. That I remember very vividly, because each one of us – at that time there were not so many – had to read the chapter and share his enthusiasm.

I knew James before because I had read about him and pragmatism and the rest of it. So for me this was rather natural; I had already noticed that Bohr's attitude was pragmatic.

DP *I seem to remember some anecdotes about investigating the mind or thoughts, or thought examining itself, that Bohr used to tell his students.*

At that time, Bohr thought very deeply about the expression of thought, that is, how we express thoughts, and how we use words. This is very characteristic, because it pervaded all his thinking in physics also. He liked W. Gibbs, because Gibbs started from a word of the common language, e.g. 'temperature,' which had a clear observational definition, and tried to connect it with an atomistic picture. That is an approach that Bohr liked, and when he thought about words and the way in which we use words in order to express our thoughts, he noticed that we use the same word in two very different meanings. We use a word to express our affections or emotions, e.g. we use the word *anger* when speaking of ourselves when we feel angry, also in describing a state of consciousness; but I can also say 'you look angry,' and there we use the word to describe not a state of consciousness, but a state of bodily activity and behaviour: the kind of behaviour which expresses anger. He noticed that this creates the risk of ambiguity, if you confuse the two; but a necessary ambiguity, because it is the only way for us to communicate our emotions; I can only know that you are angry if you tell me, or if you behave in a certain way which I interpret and identify with my state of consciousness concerning anger. It is a very deep feature of human language that it contains this ambiguity.

At that time, Bohr used a mathematical analogy with many-valued functions, a logarithmic function, for instance, which has a singular point at the origin. If you follow a path, as long as you do not go around the origin, you have a continuous variation, and you remain in the same sheet of the Riemann surface. But if you go around the origin, you reach another sheet of the Riemann surface, corresponding to a different set of values of the logarithmic function. So he said: when we discuss the state of behaviour, we must remain on the behaviour surface, and be careful not to spring over to the surface of consciousness, which is a different one. So, each word is a singularity, or is connected with a singularity, in our way of understanding existence.

PB *It strikes me that what is implied here is the subject/object dichotomy.*

That's it. This complementarity arises when we are the object of our own observation. We are at once subjects and objects. That is the very peculiar characteristic of consciousness. It is also connected with the problem of freedom of will. All the discussions about freedom of will are generally spoiled by this confusion: that will is the feeling that you have a free choice between different possibilities at the time you make a decision, but that happens on a different plane, and the concept of liberty is then no more applicable.

PB *So, in fact, Bohr's ideas in physics really had roots in philosophical perception.*

Yes. He was well prepared to recognize the physics. A situation like that is completely alien to any Kantian attitude of mind.

PB *Yet Heisenberg, even though a Kantian and perhaps even a Platonist, was able to understand, and was also ready from his point of view for the problems in physics at that time.*

That's all to the honour of Heisenberg, that he did understand Bohr at that stage, though not without a struggle. Heisenberg discovered the uncertainty relations; the background was the following.

They felt very strongly that, although quantum mechanics in the matrix formulation of Heisenberg was a complete theory, a complete logical scheme, it still did not provide a ready explanation of aperiodic phenomena. It was essentially a theory constructed for periodic systems or multiperiodic systems. That is a point that Pauli insisted upon. And in fact Schrödinger, who came out of the blue to Copenhagen (I mean to the body of knowledge in Copenhagen), provided the answer unwittingly. He had quite different ideas. He thought that he had destroyed quantum mechanics, but Pauli was quick to see, and Bohr too then, that Schrödinger's formalism provided just the way to describe aperiodic phenomena. In fact the first application of quantum mechanics that Bohr made was to the study of collisions. The two schemes were equivalent. So that was the situation.

Heisenberg (it is very curious) did not recognize the situation for a while, because he was rather stubborn and he said 'my formalism is complete, nobody denies that, and therefore *it* must contain the answer to the question of what is observable and what is not.' He believed that he had put into his formalism only observable things and discarded things that were not observable, like the orbits of the electrons and so on.

But Pauli said: 'It is not true that orbits are not observable. The orbit of the moon is observable. So there is something missing in our understanding of what is observable and what is not.'

Then Heisenberg remembered a conversation he had with Einstein in which he tried to explain his theory of observables – that he had put only observables in his theory. To this Einstein retorted: what is observable or not is not for *us* to decide, but for the theory! So when he was confronted with this problem, he remembered that remark of Einstein, and by concentrating on it, he discovered that the answer given by quantum mechanics to the question 'what is observable and what is not?' is contained in the commutation rules from which we derive the uncertainty relations. They give the reciprocal limitation on the kind of things that one can observe.

When he had the commutation rules, he thought that he had solved the whole problem. But Bohr was not satisfied. Bohr was of course very much impressed by the uncertainty relations and he saw quite clearly that they provided essentially the answer to the problem. But it was not yet formulated with sufficient precision.

Heisenberg had tried to illustrate the meaning of the uncertainty relations by a famous microscope experiment, his gamma-ray microscope. He got the idea from a conversation with a colleague, while still a student. His friend asked: 'How could we see an electron?' – more or less as a joke. His argument, when he remembered this, was that if we look at an electron with gamma rays, then, by the Compton effect, the electron is scattered in a certain direction. Therefore, when we look at it, we disturb its momentum, we lose the momentum: that was one illustration that he gave.

Bohr seized upon that, because he saw that it was quite wrong. Bohr admitted that it was true that the electron gets a new momentum from the Compton effect, but he said that we can calculate the change of momentum, and therefore correct for it. So it is *not* something which we cannot know or observe.

DP *Also, Heisenberg's argument presupposed the existence of the electron with a very precise momentum and position, which was disturbed by the observation.*

Yes. Bohr immediately rejected that view. He showed by closer analysis of the process that the uncertainty in the determination of the position was due to the angular aperture of the beam which was necessary to form an image, and that it was just this angular aperture which caused an uncertainty in the direction in which the electron would be emitted. That difference was the uncertainty in the momentum, and everything came out all right.

DP *At this point, it's not true to say the electron had an orbit or a path, in the classical sense of the words, which was implied in Heisenberg's work?*

No, certainly not. It was an instantaneous state of affairs in which neither at the beginning nor at the end could you see a definite position or momentum. This led Bohr to develop his analysis. He saw that the uncertainty relations implied two ways of looking at atomic objects which were mutually exclusive, when you pushed the idealization to the extreme.

DP *This is analogous to the remark you made earlier about discussing thoughts.*

Yes, he saw again the same mutual exclusiveness of points of view, which were both, of course, necessary. There was no question of eliminating one of them.

DP *It would not be correct then to say that the electron had a path.*

No. The first implication is that we cannot use mechanical and kinematical concepts as attributes of atomic objects. They express a relationship between the atomic object and a certain apparatus which we construct in such a way that the indication of the apparatus expresses, or defines, the concept in question. We have an apparatus from which we deduce what we call the position; we have another apparatus from which we can derive what we call the momentum. We can apply either apparatus to the object; that is our decision. We get the response. But if we have made it with one apparatus, we lose the possibility of controlling the complementary aspect.

DP *More recently, Heisenberg speaks about potentialities. Again, that isn't the same as Bohr's interpretation, is it?*

When you use such a vague word as *potentiality*, you can give it whatever meaning you like. The wave function or the state vector, whatever you call it, may be said to contain an infinity of potential answers to the question. Once you have made a measurement, let us say of position, then you get the wave function which is localized, which is a wave packet containing many values of the momentum, if you analyse it. Here one can use the word *potentiality*. Bohr was never acquainted with this idea of Heisenberg, but I can guess the way he would have taken it. He would have said: 'Well, that's a word, "potentiality"! If it is useful, all right, let us use it.' But I, personally, don't see this particular use.

DP *When we were talking to Heisenberg recently, I made the same point: that I felt this was not quite the inference Bohr had in mind. Heisenberg said that he felt he and Bohr were in complete agreement. I think you're implying that you don't feel that this is true.*

Well, no. I can understand Heisenberg. He said the same to me once when we discussed the philosophical background. Because I have good relations with Heisenberg I could tell him that he was an idealist, and that I did not like that attitude. And he said: 'Yes, yes, I see; I can understand that you have a different attitude, but the interesting thing for me is that on physics we agree completely.' That is typical, I think, of the idealist. He tries to make a distinction between the way in which he behaves when he is a physicist, or biologist, or whatever, and what he talks about when he behaves as a philosopher. Pasteur used to say that when he entered his laboratory he left his religious faith in the cloakroom. I don't think it is a reasonable attitude at all. But Heisenberg has a right to say, according to his own attitude, that he felt in complete agreement with Bohr. I can understand that. The disagreement only starts when Heisenberg begins to talk about Plato and having rediscovered numerical relationships, fundamental symmetries, and so on.

DP *Bohm has told me that nobody really understood Bohr's mind. It is a most subtle thing. The only person who could really tell you would be Rosenfeld.*

I should say that Heisenberg could do it much better. But one who understood Bohr fully and deeply was Pauli. It is again a case of not separating the general, let us call it philosophical, from the scientific attitude. You cannot understand Bohr if you try to judge, or to analyse what he says, while projecting on to his statements postulates taken from Kantian philosophy, from the idea that things have attributes. That creates a sort of barrier between what you try to read and what Bohr wanted to express.

Bohr's approach always was to say: here we have a situation which is given to us as observers, that is, as beings reacting with the universe. We have developed what is called ordinary language, which is a system of concepts by which we describe our direct observations. It is perhaps refined in physics, in microscopic physics, in classical physics, by quantitative denominations and so on. Fundamentally it is always a system for the description of our perceptions and our reactions – in general, our experience. His attitude was to consider those things as given and therefore not to be discussed. Or, at least, their discussion was another matter; it was the job of the psychologists to analyse methods of perception and the function of the senses, etc. But the job of the physicist was to start from this given experience, this given knowledge, and then to organize it into a coherent whole by using logic, since logic also is given. After all, logic is the way in which we connect various statements according to rules which are such that the conclusion is inescapable when we apply the rules.

DP *Did he mean logic in the sense of a set of rules which was observed to work in the classical world, or was it logic in the sense of something to do with mental operations?*

It is, of course, a mental operation; we are also part of the world. But there had been several attempts by J. von Neumann and others to say that quantum mechanics necessitated a new logic. Bohr was always very much against such propositions.

He considered logic also as a part of experience. He was influenced in that by his brother, Harald, the mathematician, who was at that time very much involved with the famous quarrel among mathematicians, the formalists like Russell against L.E.J. Brouwer and the intuitionists. He did not mention Brouwer, but he was certainly very much against Russell.

He favoured the intuitionists, although I would not make too definite a statement about that. Anyway, it is again a pragmatic attitude, towards logic and mathematics, just as towards our physical experience or any experience whatsoever.

In order to understand complementarity, you must first put yourself at that starting-point; otherwise you miss the point. If you are a strict logician, you will say: if it is mutually exclusive, then one of the two is false, one is right. That is obviously not the case.

PB *Historically, quantum physics comes after classical physics, and we have had several hundred years of Newton and Galileo. That's quite a long time for the language to absorb all the implications and fundamental meanings of classical science. The transition from Greek to Newtonian physics was also difficult, and required a great change in language; perhaps we could even predict eventually a quantum type language, in which such concepts as complementarity may disappear, or may be seen as something deeper.*

Bohr once told me that he hoped that after a while, when people get used to it, as you say, complementarity would be something quite natural, taught in schools, etc.

DP *I think Bohr had a rigorous view of language – that it would never be possible to talk about experience other than with the language of the classical world, that it would never be possible to have an understanding of quantum mechanics with a quantum mechanical language.*

In a sense, it is simply a question of scale. We are macroscopic objects and therefore our only approach to atoms is by the intermediary of microscopic instruments. The eye can, at the limit, perceive two or three quanta, but that's an extreme case, and not the normal way in which we use our eyes.

DP *So that language puts a barrier on any deeper, any further understanding of quantum mechanics?*

I would not say that. It makes the understanding non-trivial and different from the understanding of macroscopic phenomena. Bohr's aim – and I think he has attained it by introducing the idea of complementarity – was to make a full understanding, in the sense of description, of the behaviour of atoms possible for us.

DP *Would it be true to say that Bohr actually put a limitation on the questions that we may ask, as a result of language?*

No. His point was that there was no limitation to our possibility of describing the behaviour of atoms, provided we used the language at our disposal (the only language that is at our disposal), with due precautions which were indicated by the concept of complementarity and the interpretations.

DP *When Einstein attempted to give an objective interpretation of the wave function, Bohr more or less took this as a limit on the sort of questions one can ask, did he not?*

Yes, surely. Complementarity implies that there are certain questions which become meaningless. For instance, 'what is the position and the momentum of a given particle?' – that is a meaningless question. But we know that it is meaningless beforehand; the theory tells us.

PB *Is there a distinction between a meaningless question and an unanswerable one?*

If it is meaningless, it is also unanswerable. But the fact that we cannot answer it does not imply any restriction upon our possibilities of accounting for all possible experience that we can have with atomic objects. Our experience of atomic objects is naturally limited by the difference of scale. There are only certain experiments that we can make. We cannot see an electron between our fingers of course.

PB *I was thinking of purely statistical questions, where for some reason it becomes meaningless to try to describe the behaviour of an individual in a very large collection. That is, in a sense, an unanswerable question, and it's also meaningless, and yet does not introduce a quantum idea.*

DP *It is not meaningless to say that you cannot describe a single particle in a classical ensemble. I think meaningless is much more precise. If a statement is meaningless, its contradiction has no meaning also. I felt that Bohr was coming to the point of saying that some statements were meaningless.*

PB *In the sense of not being falsifiable.*

He had no connection at all with K. Popper, or with positivism in general. I think it would be quite wrong to connect Bohr's attitude with positivism as it is practised by people called positivists.

DP *Would it be true to say that there are some statements which are meaningless: e.g. a square circle is a meaningless thing?*

Yes, but you see they are meaningless so far as they go against the scope of the theory. Here I must make a caveat, because we are not speaking of quantum mechanics as being the last word – that is obvious. It has limited scope; and Bohr always insisted on this. No theory is more than an idealization, good enough for a certain domain of experience; what happens beyond that is a quite different problem. Heisenberg expressed this idea once: in a certain sense, classical mechanics is a perfect theory, and an eternal truth which will never be questioned in any way, although we know that it is not correct for very large motions, or very fast motions.

But with regard to this meaninglessness: the fact that certain questions about the individuality of atomic particles are meaningless is not related, as Buckley said, to the quantum idea. But complementarity is not limited to the particular case of quantum mechanics. There is another complementarity between the direct macroscopic observation of thermodynamics on the one hand, and the atomistic description of the same system on the other. They are also complementary.

PB *Because thermodynamics does not require any detailed molecular theory.*

Right, and it is there that the lack of individuality of the particles comes in. It also comes in, of course, in quantum mechanics, and there have been endless discussions among the younger generations as to whether the wave function describes a single electron or only an ensemble. Even Einstein raised that question. For Bohr, there was never any question; it was obvious that we are talking of an ensemble. As soon as we introduce statistics, we are talking of an ensemble, because statistics is made just for that. Probability implies a comparison of many similar cases with different outcomes. So there's no question; it's no problem.

DP *It is meaningless to talk about the wave function for one electron.*

It refers to one electron put under certain conditions of observation, and that is the important point to remember – that the apparatus is part of the description.

DP *I felt from reading other people's interpretations that Bohr had almost put a limitation on what we could ask, and I consider that's not really true.*

No, that's not true. Originally, Bohr thought that there was a limitation – he used the word *resignation*, which implies that you must abandon something about causality. That was, for some time, his idea, and even at the Solvay conference in 1927, where the famous confrontation with Einstein occurred, he used that point of view. But then he realized that the lack of deterministic causality does not mean lack of causality at all, but

that a statistics is another kind of causality. Then he abandoned this misleading terminology.

Einstein played a very essential part there. He was dissatisfied with this
apparent resignation, this apparent abandonment of causality. The only
kind of causality which people inoculated with Kantian philosophy had
was deterministic causality. So, at the Solvay conference in 1927 Einstein
first tried to disprove both arguments, to find counter-examples. In the
beginning, Einstein was in fact more ingenious than Bohr, in designing
fancy *gedanken* experiments which would lead to conclusions contradicting the uncertainty relations. But Bohr learned the game very quickly, and
he refuted all Einstein's proposals.

In the end, Einstein realized that there was no such trivial contradiction
in quantum mechanics. In fact, he accepted quantum mechanics fully;
that you can see from his letters, especially from his very interesting and
revealing correspondence with Besso, which has just been published. So,
if Einstein opposed quantum mechanics, it was not at all because he did
not understand any point of it. But he said: 'Es widerspricht meinem
innersten Gefühl' – it contradicts my innermost feelings. So he put the
question in the domain of feelings, or philosophical prejudice.

PB *Wouldn't you translate it as intuition?*

Einstein did not use that word, and I don't think he would have, because
we have no intuition of how atoms are going to behave, no intuition at all
about atoms. Intuition is perceiving, in a single act, a wholeness with
many relationships, which allows one to see relationships that others do
not see.

I think intuition is actually a mental operation in the same way that
logic is. It is a kind of short-cut that you can allow yourself when you see a
whole network of logical relations. Then you have to work it out carefully
to see that you have not missed anything. Intuition is a logical operation.
Some people speak of having an intuition of how an electron will behave
in certain circumstances, but that is a very abstract kind of intuition. It's
not that they visualize the atoms in any way, but that they have a formal
intuition of the workings of the differential equations.

PB *Would you say that we have no intuition of atoms, because of the nature of
our language, which is a language of everyday objects?*

Yes. Atoms are not part of everyday language; they are connected by
specific definitions with concepts of ordinary language.

PB *By the self-consistent approach which you and Professor Prigogine and
others have taken, with regard to relations between dynamics and thermo-*

dynamics, dissipative structures and biological systems, you are implying, in a sense, that all you can ever get is a self-consistent view of the world, rather than an Einsteinian one, which is almost a divine one. So it does connect up, doesn't it? Could you talk a little about your recent work in this field?

What we try to do there is just to develop and to express in a precise formalism this complementarity between the thermodynamic or macroscopic aspect and the atomic one.

PB *You have introduced the observer in the loop?*

Yes, surely. This is not the first time that this connection has been attempted, but one has always done it by introducing brute force, let us say, a statistical element, which is called 'mixing' in the jargon. It comes from Gibbs's analogy – mixing milk and coffee and getting a homogeneous mixture, whereas one knows that the molecules are not at all homogeneously distributed. So the homogeneous aspect is macroscopic because we renounce a more detailed localization of the molecules, but only consider them from a distance, so to speak, and eliminate most of the parameters assigned to the individual molecules.

Now, that is a purely classical way of speaking, and it was good enough for classical statistics. But translating this into quantum theory is another thing. Von Neumann had tried it, and, one must say, had not actually succeeded. Localizing a quantum particle and introducing a momentum distribution at the same time – this gets you into conflict with the uncertainty relations. Of course, there are tricks. E. Wigner has introduced a very neat and elegant trick to get around that, but it is just a trick and does not give any satisfying solution.

Now, Prigogine's idea was to consider infinite systems so as to get rid of the periodicity which is inherent in the mechanical behaviour of finite systems. If it is infinite, then the period becomes infinite also. So you consider only a stretch, so to speak, a necessarily finite stretch of an evolution, which has no end in the finite. You can even push away the beginning to minus infinity, if you like. Then, if you try to determine the asymptotic behaviour of such an infinite system, you *do* find that the phase relations are automatically eliminated from the asymptotic density function without any necessity of explicitly introducing a statistical element. The statistical element is there, of course, but it is contained, inherently, in the whole description, so that the mixing is produced by the system itself and not by any imposed coarse-graining. The coarse-graining is inherent in the behaviour of the system itself. This puts the complementarity on a similar footing to the complementarity of quantum mechanics, where also it is not our doing that there is this complementarity between position and momentum; it is a consequence of the existence of the quantum of action, the fact that the atoms are not able to interchange action except in multiples of a unit.

PB *How does this complementarity tie in, now, with irreversibility and time flow? For instance, Weizsäcker's work introduces time on a very fundamental level. This seems very new in physics.*

Yes, we realize that the mixing, which in Gibbs's conception was the element producing this irreversible behaviour, occurs as a consequence of the dynamics of the system. The irreversibility which is a consequence of this mixing is inherent in the behaviour of the system – even in purely dynamical systems, in spite of the inherent reversibility of the micro-scopic behaviour.

PB *Because there are thresholds where order is possible?*

Yes.

DP *Is it really possible to talk about microscopic behaviour without at the same time specifying some microscopic system to which it refers?*

That's a very involved question. When we describe atomic behaviour, we use a classical language, even if, or especially if, we speak of quantal behaviour of the atomic system. In this part of the description, the rever-sibility of time is included. But then, by this asymptotic process, we de-stroy the invariance of the equation with respect to time-reversal. So the microscopic description that we obtain no longer has the character of re-versibility in time. That is done by what we call a projection. That is a most technical detail, and a very abstract thing. It corresponds (to try to put it in ordinary language) to the fact that we eliminate most of the parameters which describe the behaviour of the atomic system. We only keep those that we decide are directly observable. I say 'we decide,' be-cause, after all, we can, if we like, observe an atom by using a gamma-ray microscope, or that kind of apparatus. That is perfectly permissible, in the logical sense, even though we cannot build such instruments. But in bubble chambers, at Geneva and other places, there is apparatus which actually shows us individual atomic processes.

PB *It's interesting that we interpret the microscopic bubbles in terms of a path.*

Yes, but we understand how this apparatus works.

PB *At least we can observe some atomic events.*

Processes, yes. A bubble chamber picture is terribly complicated. Every bubble is a single experiment. But, then, thermodynamics is another mode of description that we have found useful (I'm becoming pragmatic again), and which consists in the elimination of most of the parameters and the retaining of only a sort of global effect which we call pressure, temperature, etc., which are averages over many atomic processes. Then we see that by coming from the atomic description to that new description by this elimination, we have also eliminated the invariance with respect to time-reversal.

DP *This break with symmetries, though, is a characteristic of the classical world in many cases. Do you think that this is always true, that somehow the symmetry has been broken by going from a system with a very large number of variables to one with just a few classical variables?*

That is certainly true in *this* case; I don't know how general it is. It is reasonable to expect that when we lose symmetries, we eliminate characteristics.

DP *Heisenberg has a theory in which he has a fundamental symmetrical law and the world breaks the symmetry of the law. In fact, this may be going to asymptotic states in which the symmetries, or the very fundamental particles, are all broken.*

Yes, that may very well be. We are not that far yet.

DP *Do you think, in this way, that it may be possible to have a theory of relativity, of generalizing the gravitation by taking an asymptotic limit?*

It may be that it will turn out like that, but we have no microscopic theories, so we can't say anything about that.

The kind of irreversibility we get depends on the questions we ask. Usually we are interested in the future, and therefore we have this aspect of dissipation, the reversibility getting worse and worse. But we are also perfectly able to look to the past and make retrodictions, which are also statistical of course. We can ask what the probability is that the present microscopic state of the system has arisen from a given configuration. We can also formulate the theory so as to get retrodiction.

DP *Weizsäcker has said it is fundamentally not correct to use the term* probability *in this sense, that is, to speak of retrodictive probability.*

PB *They're all future-oriented, because, in a sense, you are putting yourself back in that supposed initial condition. In its formal sense of use, he's right, obviously.*

Yes. He formulated that idea at a very early stage, in 1940 in a paper where he mentions that he received his inspiration from Gibbs. Gibbs says, if I may try to paraphrase his long and very complicated sentence, that in trying to make retrodictions about previous events, we are really able to disregard our knowledge of the probabilities of anterior events that have influenced those that we contemplate. I meant to mention only very trivial retrodictions, which we put ourselves in; the unrealistic situation of knowing nothing at all about the past of the system, which is of course never the case.

DP *There is a search for fundamental particles – partons, quarks – continually reinterpreting theory in terms of more particles. I was wondering to what extent this is a failure to get rid of classical ideas and the notion of a particle, even a psychological failure.*

I think it is. I think the people doing the latest things in elementary particles are rather crude in their thinking. They are in danger of getting into a situation of infinite regress. If you introduce quarks, which must be very tightly bound together, what is then the field, or whatever, that binds them? And so you can go on indefinitely. So I think this is not a fruitful way to look at things.

PB *What about fundamental symmetries?*

The fundamental symmetries give a very strong indication that those things that we call elementary particles are actually structures consisting of elements which can arrange themselves in different ways. But that does not mean that those elements can be compared to the crude idea that we have of particles bound together by forces of another kind. That would lead us to an impossible problem. But they may stick together in the way that Heisenberg contemplates, by self-interaction.

David Joseph Bohm

David Bohm is Professor of Theoretical Physics at Birkbeck College, the University of London. Born in Wilkes-Barre, Pennsylvania, in 1916, he gave some hint of his future scientific vocation when he displayed a childhood interest in mechanical devices and planned to make his fortune as a boy inventor. About this time he had sensations of the 'interconnectedness' of the world, a revelation which appears to have influenced his later thinking.

Bohm studied physics with Ernest Oppenheimer and, as a young research physicist, voiced his concerns over the foundations of scientific theories to Albert Einstein at Princeton. Bohm's early research on electron plasmas in metals is still considered a significant contribution to the theory of the solid state, but he was soon to leave 'conventional' research in favour of an investigation into quantum and relativity theories and the possibility of their unification.

Bohm has not yet been successful in formulating a more general theory of physics and it could be said that his greatest contribution has been in causing physicists to re-examine what it is they are doing and to question the nature of their theories and their scientific methodology.

Recently Bohm has become interested in education, its effects upon the developing individual, and the future of society. He has therefore become actively engaged in an educational experiment at Brockwood Park, England.

Bohm's passions are for conversation – he is an animated talker – and walking. A colleague who is fortunate enough to start David Bohm on a train of thought may find himself involved in a cross-country hike-cum-discussion which will last for the remainder of the day! The following conversation with David Peat took place in London and for once did not involve any walking.

Most of the physicists with whom we have had conversations have tended to accept quantum mechanics as it is. They are trying to extend the formalism a little, either to unify it with relativity, or to attempt to provide an explanation for the elementary particles. I take it that you are not really satisfied with this approach.

Perhaps I should go back into the history of how my ideas came about. When I studied quantum mechanics I was fascinated with it. I felt it was a very deep, important study, but I didn't really understand it. Eventually I taught a course on the subject, and wrote a book on it, to try to understand it. After finishing my book [*Quantum Theory*, Prentice-Hall, 1951], I considered the matter again, and I felt that I still did not understand it. At that time, I began to think of different ideas than the usually accepted ones. I sent copies of the book to various physicists, including Einstein, who expressed interest in it, and we had some discussions. I think we agreed that one couldn't really understand what quantum mechanics was about. I also talked with Oppenheimer, but he was never critical enough to make possible a discussion at the level I would have liked. I sent my book to Pauli, who liked it, and also to Niels Bohr, but I received no comments from him.

Since I can't remember exactly how I thought at that time, I'll try to say what I now think the difficulties are. This is probably similar in essence to what I felt then. Any theoretical science has four aspects. These are: *insight*, to perceive the structure of new ideas; *imagination*, which projects a mental image of the whole idea, not only a visual image, but a feeling for it; *reasoning*, to work out the consequences logically; and, finally, *calculation*, to get numbers that make possible precise tests with experiment. Evidently all four were present in physics until quantum mechanics came in. In quantum mechanics people discovered that they could find no way of imagining the meaning of the theory. This was brought out most clearly and consistently by Niels Bohr. I'm not sure that any other physicist really understands exactly what Bohr meant to say, but I don't think we can discuss that here.

It is rather widely believed nowadays that science, at least physics, does not give much scope to imagination. Various imaginative pictures are used, like 'wave' and 'particles,' but they are in no sense regarded as a real description of what we are talking about. They are merely aids to calculation; we deploy our imaginative pictures so that we can calculate more efficiently.

What do you mean by understanding?

I mean to grasp the whole thing, to get a feeling for the whole thing. If I become proficient in calculating results, I don't feel that I necessarily understand what it's about. By way of example, I might make a comparison with the Newtonian epoch. Let us say that Newton developed a calculus, and became very proficient at it. Every time you have the power x to the nth, you would replace it by nx^{n-1}, and you can go through all sorts of operations until you can finally say that you are proficient at working out these operations and can get numbers. Meanwhile, some other experimental physicist is proficient at manipulating his telescope, and he gets other numbers. If the two numbers agree, then everybody's happy. When the numbers disagree, they aren't happy and try again. That would have been the way quantum mechanics was done. I don't think Newton thought that way. He had some sort of imaginative overview of the whole meaning of the thing, of the universe.

Do you think this was why Newton was very concerned about gravitation, because he didn't really understand it?

That's right. For him it was only a means of calculating, and he was not satisfied. Modern physicists would say that they don't care, that's all a physicist wants to do. That is the change of attitude. I recall Feynman writing that imagination was the most important thing – and he is an imaginative fellow – but finally it always works out that the calculation is the main thing. I regard calculation as significant only to test the other aspects of physics. In itself, I regard it as rather insignificant. I don't think that the things physicists calculate are very interesting – e.g. how many Geiger counters are going to click; how many spots will appear on a photographic plate.

So it's really a test of the consistency of your own understanding.

Yes, and of the factuality of it also. Is it a real understanding? If you have an imaginative insight, you want to be sure it's not just imagination, you have to see that it's factual.

Wasn't it Goethe who attempted to postulate a physics based upon our everyday experience, rather than on making experiments and creating artificial situations?

When Roger Bacon originally suggested the form of modern science, he suggested that experience should play the key part in testing. Before that time people thought that Aristotle was the authority for what was true and, if you disagreed with Aristotle, you must be wrong. So it was a tremendously revolutionary idea to say that experience should be the test. This was later elaborated to say that one should try to arrange special experiences which are very simple. Ordinary experience is so complicated that it's very difficult to see just what it is testing. Then experiments were elaborated. This is a very powerful method but, at the same time, dangerous, because the experiments are developed on the basis of the theory; they are set up to answer the sort of questions that a certain theory asks. When experimental equipment was very cheap and simple, it didn't matter if one experiment or theory did not work out, because another theory could be considered, and one could try another experiment. But *now* it takes years to produce a big machine; it requires the cooperative work of many people and millions of dollars. People feel that once you have invested in this machinery, you had better use it. Theorists then feel impelled to develop theories that will raise questions that can be answered by this particular equipment, which in it's turn was set up to answer questions due to the previous theory. The result is that the experimental method, as it has developed, may tend to introduce a very conservative factor into physics whereas, in the beginning, it was quite radical and revolutionary.

Would you say that this is true of the particle accelerators, that they are perpetuating a fragmentary view of nature?

I think a lot of people are questioning particle accelerators. The very fact that they are not supported now to the extent that they once were indicates that many physicists feel that they are not likely to produce the results that were expected. It was discovered by E. Rutherford that if you bombard atoms with alpha-particles, you can learn quite a bit about them. But that depends upon the idea that there is something stable about the atom, which remains while you are bombarding it. Now we are using such high energies that we literally disrupt everything and create all sorts of new things.

We could compare this to trying to study the structure of cities by bombarding them with higher and higher explosives and studying the fragments. If you bombard them with light, which doesn't destroy the cities, you learn something. If you use some sort of very fine shot, you might learn something, but as you raise the energy, you learn less and less rather than more and more.

You said that there was difficulty in understanding quantum mechanics.

Yes. I think that the difficulty is that we have no way of understanding what is actually happening, or what I call *the actual fact*. If I may paraphrase Bohr, we have only the phenomena, i.e. the observed phenomena, which are essentially classical in their description. Ordinary classical phenomena – the observation of a dot or a click – were previously understood to signify information about particles, and the particles were independent of these phenomena. Now, if you analyse the Heisenberg microscope experiment, you come to the conclusion that the experiment cannot give you unambiguous information about the structures you are supposed to be observing. Therefore, there is no clear way of considering the unknown reality which is responsible for the experimental result.

Wouldn't Bohr have said that this is a fundamental property of the world?

In effect he did say that. I don't think that he ever said it directly, but it was implied. But if he said that it is fundamental, then I ask: how does he know it's fundamental? It's only fundamental as long as the present theory works, and there are many ways in which it doesn't work, as we know. We certainly just can't accept it on authority that it is fundamental. We don't have Aristotle to tell us what is fundamental and what is not. Neither can our experiments tell us what is fundamental and what is not, because, as I've said, our experiments answer only the questions that we have already asked.

What about Bohr's view of language itself?

I would ask again: how does Bohr know that? I think the nature of language is even more unknown than the nature of particles. Bohr said that we are suspended in language and we literally don't know which way is up and which way is down; yet we are compelled to use language.

Our language has certain concepts in it and he believed that our language is committed to the concepts of classical physics, at least ultimately. That is, the ordinary ideas of place and time, and object and substance and matter, eventually, when refined, lead to the classical concepts of particles with certain positions and momentum. Bohr believed that the only way to get unambiguous communication is through classical concepts, and he takes it to be the task of physics to have unambiguous communications. But, contrary to Bohr, I say that physics is not primarily concerned with unambiguous communications; rather that *all* concepts are ambiguous, and that there are certain unambiguous abstractions that can be made from our ambiguous concepts. Those are the things that we use for tests. I think people get it upside down when they say that the unambiguous is the reality and the ambiguous is merely uncertainty about what is really unambiguous. Let's turn it around the other way: the ambiguous is the reality, and the unambiguous is merely a very special case of it, where we finally manage to pin down some very special aspect.

In his early works Wittgenstein said that words were justified by their relationship to facts in the world, but later he said that it was in their use. Perhaps what Bohr said was too limited, and language is much more subtle than he believed.

First of all, you can't discuss language apart from thought. Language is only noises unless it is expressing thought. I don't think anyone would presume to say that he knows the structure of thought. Not only is it unknown, but he would get into a terrible tangle, because of the very thought with which he is thinking about that structure: does he know *that*? Isn't there a danger that he is projecting some idea which he has in calling it the objective structure of thought? That's just the same problem as in machines. Machines have been built up in such a way that they lead us to ask only certain questions. If you have a theory of the structure of thought, you will project it into your thought and say that's what thought is. Then you will ask only questions about thought which are in your theory, and your thought will only answer the questions that you ask. So you are caught.

Are you saying that this is a limitation of our knowing?

I'm saying that any idea which attempts to state that we know the structure of thought, or the structure of language, is suspect in my view. For example, N. Chomsky has stated that the structure of language is based, as I understand it, on our brain structure. He thinks he can connect it up. This may be insightful, but if he thinks he knows the ultimate structure of language, I think there is an extremely dangerous possibility of self-deception.

The structure of language and the structure of thought are essentially one, inseparable. There are thoughts that go beyond language, but you cannot discuss the structure of language apart from the structure of the thought that language expresses. That is infinitely subtle, and I think Bohr might even have agreed with that. But Bohr made a still more subtle point – he was an extremely subtle person and very difficult to understand. Bohr said that he understood how subtle language is, but that physics is confined to dealing with unambiguous concepts, whose meaning cannot be doubted. Now I want to question that. Art is a field where ambiguous concepts are obviously the rule; you don't expect an image in art to definitely mean exactly this or that. But people think that physics means exactly such and such – at least that's the way that Bohr put it. I don't think that physics does mean that. Physics is a form of insight and as such it's a form of art. Every fundamental theory is an art-form in my view, and we can see how this art-form fits our general experience. No art-form fits it perfectly, so we go from one to another.

Classical physics led us to the ideal that we have a perfect correspondence between concept and fact, and thus no ambiguity. But when people study even classical physics carefully, they find contradictions. Zeno's paradox is a case in point. The most fundamental classical concept is an object moving through space, like a particle. As Zeno analysed it, a particle is in a certain position, then it's in another one, and another, and so on; while it's in a certain position, it cannot be moving; when it's moving, it cannot be in a certain position. The concept of motion involves an essential ambiguity in the position. In fact, in our mathematics, if you take a certain point, according to the theory of continuity of a line there is no next point, it is ambiguous. But it follows that the *present* point is also ambiguous. What do you mean by the present moment? That's ambiguous because it's too fast. If you try to point to what it means, you don't get one moment, but you get some ambiguity as to exactly what it means.

You're saying that physics has aspects of an art-form, so what criteria do you use for working in physics?

What criteria do you use in art? People have never been able to answer that question. I don't think you can answer it in physics. People are looking for complete security by saying they know a certain criterion by which they can judge what is good physics. But any attempt to make that criterion will just kill physics, because almost any new idea is bound to disagree with that criterion. The word *art* in Latin is based on a word meaning 'to fit, that which fits, that which is in harmony.' Ultimately, we have to see the harmony or fitting of our thoughts and our broader experience. If you have a preconceived idea of what constitutes fitting, then your mind is blocked. You may need something different.

You're stressing the idea that science is a human activity.

It's a creative activity.

... and a personal activity.

Well, it's both personal and collective. But I would rather emphasize that it's creative and not mechanical. Something new has to be created. If you have a fixed criterion of what fits, you cannot create something new, because you have to create something that fits in with your old idea. If we say 'science is "X," science is something that fits a certain idea' – namely, what people have thought science is – then that limits what we can think.

If we have the concept of what fits, we're limiting ourselves. Then how do we carry out this activity in our lives?

Is that a good question? When you ask the question 'how do we do it?' you're asking for a plan of how to go about things. This denies creativity. It is like saying 'how can I become a great and creative artist?' Can there be a technique, or a plan, or a criterion?

But you wouldn't presumably go so far as to say that a critical analysis is not involved in the way your life is carried out?

Even that has to be creative. You can't take a fixed form of analysis. Any attempt to determine this thing beforehand is arbitrary. You are going to choose the criterion you prefer or enjoy, or the one that society enjoys or prefers.

But you must have some criterion. You're claiming that people working with accelerators are doing things that don't fit.

I haven't stated a criterion. I'm just saying that if you look, you'll see that it doesn't fit. How do you tell that there is a contradiction? Is there a rule for recognizing contradiction? It's the same as seeing that a picture is disharmonious or that a piece of music is not in harmony. What was once called disharmony in music later was called harmony. You can't fix the thing.

So the notion of attempting to fit a picture onto reality is completely alien to what you're saying – I mean the notion of a reality which exists independent of man.

There is a reality which is beyond man, and includes man, but this is unknown. A man has certain ideas which dispose him to act in a certain way. If this action is harmonious, then he regards these ideas as correct. Our thought disposes us to act in a certain way. The word *dispose* means 'to arrange,' as a commander disposing his forces. If they are wrongly disposed, he will get into trouble, the worst trouble being the disposition of one-half of his forces against the other; that's a contradiction. For example, you are walking down the stairs in the dark and your body is disposed to expect another stair, but it happens to be flat. The whole movement is disorganized; it is not in harmony. Then suddenly you have the thought that this is flat and the entire disposition changes. I think that's the way all our thought works, including scientific thought. A certain way of thinking disposes us to act, in the laboratory or elsewhere, in a certain way. As long as we can find some general harmony in this action, we go on with it. When we find disharmony, we hold back and begin to look for another form of thought.

Would you say then that physics today just doesn't fit, is not in harmony?

It's not in harmony. Quantum mechanics has no imaginative conception. If you are satisfied to say that physics is nothing but operating a formalism to get results, and operating equipment to get results, in order to obtain results which agree, all right. But if you say that physics aims to understand what's happening imaginatively, then I don't think that it's doing that. Neither relativity nor quantum theory is clear. And the relation between relativity and quantum theory is even less clear.

Was this lack of fitting, this disharmony, true even before relativity, in the last century?

There was always trouble with classical physics but it was never quite so dramatic. There have been problems such as 'is there an ether?' or 'is there absolute motion?' Newton had the idea of absolute space, but it wasn't clear what he meant by it. He said the fixed stars are the frame of absolute space, but why should they be?

Have the experiments of quantum mechanics and relativity actually exposed some long-term error in our way of thinking about the world?

I don't think it was the experiments, but the theories themselves. The insight in the theories exposed an inadequacy in our way of thinking. It implied that we should have gone further to develop new ways of thinking, but this has not been done. As Bohr said, we have only classical concepts, they are the only unambiguous concepts. I believe that we cannot understand movement if we insist on unambiguous concepts.

Did the fundamental experiments of quantum mechanics really show the error of the notion of man confronting nature as a separate object?

I think they do, but it's a very subtle thing to analyse. Bohr has given the most consistent analysis, but it's very hard either to understand or to express what Bohr meant. Generally the position is this. In classical physics, we say the world is made of separate objects, each a separate substance in mechanical interaction. The observing equipment is one of these objects, and therefore can be influenced by the other objects which it is observing. Evidently, you can maintain the separation of the observing equipment from the object observed and, in turn, the observing human being from the equipment, and so on. In quantum mechanics, one sees that the process by which these different things would interact cannot itself be analysed in detail. It is whole and indivisible. You cannot make a separation between the observing instrument and what is observed. For example, you are looking at this table; the form of this table has been built up by your experience which you are projecting into the table. Is the table you, or is it something separate from you? You appreciate it as a table with a certain form and a certain subsistence, but that form and that subsistence are as much you as the table. If you probed it with a very high energy machine, say with neutrinos, they would go right through, and the table would be a vaporous nebula. So the form of the table as a solid substance, or subsistence, comes from the human brain with its own particular mode of interacting. In a sense, the observer is the observed.

Something similar must occur in physics. We probe matter with certain ideas as to what to expect, and we make instruments in accordance with those ideas. In so far as the whole procedure works and fits, we say that is what it is. But later on, we will say it is something else. We once said it was a little billiard-ball atom, and now we are saying it is something very different. The difficulty is that we see a lot of new things, but then try to explain them by particles. These particles would have to behave like waves at times; they would have to pass through barriers which are unpassable; they would have to spread out like a wave and suddenly condense; they would have to jump from one orbit to another without passing in between; and so on. They would have to do all sorts of things that particles can't do, yet we still call them particles. I think that most physicists believe that they are getting the ultimate constituent substances of the universe by discovering particles, although these particles behave in a way which would suggest that they are not that at all. By calling them particles you dispose your mind to think of them that way, in contradiction to some of the other properties that you're ascribing to them.

So it is necessary to engage in self-examination constantly if you wish to pursue science.

You have to examine your thought, which is self-examination. People generally take their thought for granted. They pick up their way of thinking in school and from their parents. They say: 'we'll examine everything else, but we don't have to examine thought, we'll just think.'

By thought do you mean something different than logic?

Logic is only part of thought. Thinking is not only logical. In fact, thinking is usually not logical. People have to go to a great deal of effort to make their thought logical.

So science is not founded purely on logic.

Many would say that science is founded on logic and experiment, but I don't think that it's just logic and experiment.

Some feel that it's possible to build up quantum mechanics from logic. Would you care to comment on that?

It depends on what you mean by logic. The root of the word *logic* is the Greek word *logos*, which is, according to the dictionary, the inner or essential thought of the thing. Many of our thoughts are just on the surface. It has also to do with the idea, or the word. But the question is: what is the relation between logic and reason? Reason is an activity. I call it perception through the mind. For example, when Newton saw universal gravitation as the reason for the behaviour of the planets, this was perception, it was not a deduction by logic from previous facts or ideas. Reason then is essentially perception through the mind. Logic is a way of trying to organize our thoughts so that they will be generally more harmonious. However, what we ordinarily mean by logic is a set of rules for organizing thoughts that are already in existence – arranging them, disposing them. But reason – perception through the mind – creates new orders of thought. Without this creation of new orders of thought, I think we wouldn't get anywhere.

But it may be possible to found an existing theory on logic or on logical structures.

You can examine its logical structure if you like, but that's not the foundation of it, you see.

Does the foundation lie in thought?

I would say that theory has no foundation. Any creative act or process has no foundation. If it has a foundation, it is not creative. If it comes from something that is already there, then it is merely the working out of what is already there, so it is not really the creation of something new.

Then what is the logical analysis of theories?

You may analyse them to see whether your ideas are clear. Very often our theoretical ideas are confused in the sense that they may point in several directions at once without our knowing it. Perhaps a logical analysis can reveal confusion; it has done so on occasion. When you see confusion, that means that you should drop the theory and try another one. If logical analysis reveals confusion then it is valuable, but I don't think that it plays a fundamental role in the theory itself.

Would you comment on the work of David Finkelstein* on logics in quantum theory?

I can't say that I fully understand the work, so I'm reluctant to comment. The ordinary logic of common sense, which, when put in mathematical form, has been called Boolean logic, implies that a proposition is either true or false. The word *proposition* is interesting. It is 'a proposal, something put forth.' If we took that literally, I think we could be much clearer. We could say that the function of thought is to propose, to put things forth. But there has to be an act of observation which *disposes*, that is, which judges between the true and the false. The judgment of a proposition should properly involve an act of observation, but in mathematics it is often an act of demonstration. This demonstration requires observation

---

*   It was intended to include in this book a discussion with David Finkelstein on his attempts to derive space-time structure and the spectrum of the elementary particles. In the course of the discussion Finkelstein was to have commented on Bohm's reactions to his work. The proposed interview did not, however, take place.

at the intellectual level, or perception. Science has generally accepted Boolean logic, but quantum mechanics has certain operators, whose value is either 0 or 1, which could be used to describe a proposition being either true or false. Quantum mechanics has sets of operators, such that you can have one set of propositions all compatible with one another, and another set of compatible propositions, but the two sets are not compatible with each other. Quantum mechanics allows us to make a model of mutually incompatible propositions in terms of sets of operators that don't commute. You also find that when you have operators that don't commute, something more is needed, namely a discussion of the context in which any particular operator is the relevant one.

This is true in ordinary reasoning also. For example, the statement that the electron is either green or not green. We consider the context of what we know about electrons and find that this makes no sense; it doesn't fit the context at all; it is not a question that should be answered. Similarly, in quantum mechanics, questions could be asked about whether an electron is in a certain position or not; in another context, whether it has a certain momentum or not; but not both questions together. Consider spin for example: in one context, when our apparatus is oriented in a z-direction, we can discuss whether the spin is up or down; in another context, when our apparatus is oriented in the x-direction, we cannot discuss that, but only whether it is right or left. The two questions cannot be relevant together.

I don't detect in Finkelstein's work any real emphasis on this context-dependence. I think that this is a weakness. If you accept context-dependence, you can see that logic is in some sense empirical, that it is not purely a question of truth. (In fact, Finkelstein has said this himself.) Whether a certain set of propositions is relevant or not depends on the context, and that has to be seen in some broader way that goes beyond logic. One of the things that is missing is some broader imaginative concept which would show us why properties depend upon the context, or why propositions depend upon the context. I have developed an idea of this which I call the *implicate, or enfolded order.*

Can you give an example of what you mean by implicate and explicate order?

Take a jar of a very viscous fluid – say glycerin – and put a drop of insoluble dye into it. There is a device that turns the whole thing slowly, stirring the mixture until it becomes grey. Then you turn it around the other way, and slowly the thread of grey dye pulls together and makes the original drop again. While the fluid was grey – i.e., the dye was all spread out – the drop of dye was still there, in some form, but it was folded up into the whole liquid, or implicate. *Implicate* means, in Latin, 'folded up' or *enfolded*. On the other hand, *explicate* order is the one that is *unfolded*. I can understand the quantum properties of these operators by considering that any phenomenon that is unfolded before us in a laboratory could be regarded as, generally speaking, folded up through all space.

Would this be analogous to illuminating a holographic plate and unfolding the image?

That's right. In the holographic plate, the information about the image is present, folded up, all through the plate. Similarly, the information that determines how our apparatus is going to behave is contained, folded up, all through space. Therefore, you no longer have the model of localized objects, which are independent substances, as the explanation of everything.

In thermodynamics, people speak of entropy as being related to disorder. Disorder, to you, would be a form of implicate order?

Yes. There is no disorder in the sense of absence of order, but rather there are different kinds of order.

So is entropy a measure of implication?

I should think that you could look at it as a certain kind of measure of implication, or 'enfoldment,' of the order. Since the enfolded does not appear obviously on the surface, you call it disorder because you don't see it. But, obviously, we can't say that anything that we don't see doesn't exist because we don't see it. I think that there is an order that we ordinarily don't see, because we are looking for something with unambiguous significance, that is, explicate. The fact that there are propositions which are not mutually compatible is a sign that the basic order is implicate. So that when one proposition is explicate, the other must be implicate, and vice versa. This would give you an imaginative understanding of why we use this logic.

Could this idea of one proposition being implicate and the other explicate be related to the uncertainty principle?

Yes. We say that all properties cannot be explicate together. But I would rather call it – as Bohr might have called it – the 'ambiguity principle,' not the uncertainty principle. The word *uncertain* implies that it exists in a definite form and that we are just not certain of it; we don't know of it. Rather, we should say that this property is not uncertain but ambiguous; that is, it has no clear meaning. We must give it a very complex representation in terms of many images.

Would you like to comment on the structure of mind in relation to implicate and explicate order?

It suggests a structure in which mind and matter are not very different. Anyone can see that our thought has this character, that a large part of it is implicit or folded up. When one part is explicit, a tremendous amount is implicit. As we talk, the words are explicit, but the whole meaning is implicit; we couldn't pin it down. This implicate order is common to mind and to matter, so it means that we have much of a parallelism between the two sides. Naturally, this will require a great deal of development. The things which are well defined and explicate have to be seen as special features of the implicate order. The underlying reality is the implicate order, and the explicate order is a very special case of the implicate order.

Would you connect this with the quantum mechanical notion of wholeness and the absence of fragmentation?

This idea of implicate and explicate order obviously involves wholeness, because, in the implicate order, everything has its origin in the totality; it is folded into the totality. Moreover, the separation of the observer and the observed is no longer basic in this view. The observer is essentially an implicate order, and so *is* the observed. Everything that is observed is really the intersection of two streams of energy: one stream which belongs to the thing observed, the other which belongs to the observer. The 'phenomena' are the result of the intersection of these two streams. Both streams come ultimately from the same total reality. There is a total reality which cannot be pinned down, it is ambiguous. It can be thought of as having many, relatively independent, streams of movement or energy. The physicist sends one stream of energy in the form of a beam of particles; the other stream is the target, which is more or less a stationary stream, moving only inwardly in the atoms; where these two intersect there arise phenomena.

The words and ideas you use have the sense of things coming into being, or time, but I think that you consider time in a very different sense?

I think that time is not the fundamental order, but a subsidiary order which man's thought has introduced. You can see that time is full of paradoxes. If we think of the past, the past is gone. We can never get hold of the past. The future is not yet, it hasn't come. We can never get hold of the future. The present is much too fast to get hold of; by the time you've said it, it's gone. So you can't grasp the past, the present, or the future. So what is time? Time is, at most, an abstraction introduced by thought. You cannot get any exact moment of time, except in your thought. If I describe something as happening in time, whatever I describe has already happened; it's only present in my memory.

But the enfolding and unfolding which take place in nature surely take place in time?

We have to think that over carefully. Let's try to see what is difficult about the concept of time. Anything that I describe is gone. Of course, it may change slowly so that it is not too different, but in fact it is gone. We may expect that it hasn't changed very much, and in some sense that is true. But when it comes to anything which is really subtle and fast, for example elementary particles in physics, or the attempt to discuss the mind, then that short interval between the past which is gone and the present which is unknown may be all important. But all physics developed thus far depends upon the assumption that it is not important; but that's only an assumption.

We say that the function of physics is to predict. Present knowledge is actually knowledge of the past. Also, we are not predicting the future as it is, but the future as we shall see it in the future, which makes the future into the past of the future. We never predict from 'what is' to 'what will be,' rather from 'what has been' to 'what will have been.' Then we make the assumption that what has been is very close to what is, and that what will have been is very close to what will be. What we often fail to realize is that the assumption depends not only on the slowness of movement, but also on the metaphysics which says that everything is made of things, like particles, which don't change very much as they move, which move on a path that can be followed, and so on.

But aren't you using 'move' in two different senses, one in moving through time, and one in the movement of the implicate and explicate order?

So far I haven't tried to define what I mean by movement. There are various ideas involved. Classical physics has the idea of the orbit as a description of movement, but the orbit is an abstraction, you never see the orbit. The past positions may be plotted on a piece of paper, but they are not seen, they are gone. They may be present in your memory, but the orbit is an abstraction which exists, as far as we know, only in somebody's mind. Zeno's paradox raises that point too, because it says that when the particle is here at a certain moment, the past position is gone. So how can you define movement as the relation between the present position and the past position? You would be trying to define a property that exists as the relation between what exists and what does not exist.

Some have said that time is the fundamental thing and movement is derived. Are you saying the opposite?

Movement is fundamental and time is an order which we derive. Movement is the fact with which we begin. You cannot specify movement unambiguously; movement cannot be given an unambiguous description. Look at your own experience. Do you ever actually see time? You never do. You see the position of a clock. You may remember time, but you never perceive time. The memory of time is a set of images which is present now, but ordered by thought. And that's a clue. You do not actually experience movement by remembering a series of positions. If you are in a moving car, you feel that you are moving; you don't say I remember that I was there, there, there. And if you project a series of positions on a screen, they are not experienced as movement until they are so close that they are no longer unambiguously separated. Then you feel movement.

Do you think there is a danger in making an analogy between movement in space and movement through time?

But I don't know what movement through time means!

Well, people involved with relativity talk in terms of a body moving through time.

That's what they say, but I don't know what it means. It's the same as for quantum mechanics, I don't understand a great deal of what is done in relativity. If I project time t as an axis, certainly I can see the track as it's drawn on a piece of paper, but the track drawn on the paper is not the movement.

I ask: What is movement through time? What is time? Time exists only in the mind. Does the particle move through the mind? I don't get it! It is almost like treating time as a substance. If I say move through London, I see what that means, but move through time? First I'd have to see what time is, and then I would see a particle moving through it! But nobody sees that.

You're saying that all these problems must be made clear before there's any chance of making progress in physics?

Before fundamental progress is made, yes. I think that the main questions to be considered are: 'what is time?' 'what is movement?' and 'what is thought?' I believe that time is entirely constructed by our way of thinking. You find time only by recalling images of what has been. Those images must be based on what is in the brain, but the 'what is' is implicit or enfolded. Remembering the past consists of unfolding this image into a series, and we say 'that's the way it was.' The future consists of unfolding it in the way we expect it to be. But movement is not experienced as moving through these images. That is merely a way by which we know something about it; through these images we can dispose our activities toward the next step.

Actually I would like to consider the notion of flow rather than movement. There is an unknown reality which can only be described as eternal flux or flow. Out of this appear various forms which can be perceived. When these forms have a certain persistence and stability, we can recognize them and we call them objects. But we must consider our attitude to these objects. Our attitude is that all objects, such as this table or this microphone, are not only forms, they are substances, and they exist independently. Therefore, the form belongs to the substance. The other attitude is to say that they are not substance, but they are *subsistence*, they have a certain stability. For instance, the vortex has a certain stability in water, but it is not an independent substance. Ordinarily, we take the view that water is the substance, but if we try to analyse water into atoms we get into trouble because of their quantum properties. So I would say that the substance cannot be pinned down in any unambiguous way at all. It is unknown. But we can abstract forms in the movement of this substance. The true substance, however, is that which determines its own form.

Are things really thought to be substantial essences?

In some sense, yes. The whole atomic theory is the idea of substantial essence. It says that every atom is a substance, and that it has a form which is the form of that substance. The world is full of independent substances, one for each atom. But that doesn't work you see. Every atom has been broken down into smaller particles and these into smaller. People call the latter particles partons. They hope that they have found the ultimate independent substance. I think we should coin a word, which I call the *ultimon*, the ultimate piece of independent substance, out of which everything is made. I think it's an illusion!

But classical physics was based on that sort of illusion and it seemed to have worked quite well. People thought that they had an understanding of nature.

Obviously it works. These forms do have subsistence and stability, so a possible explanation of this is to say that they are substances.

Yes, but at that time was there a harmony or a fitting?

All theories have this character, that there is harmony and fitting up to a point. When you push them further there isn't.

Music is an example. There was harmony and form; then it changed and we have a new harmony and a new form.

At least a search for a new harmony. Art is in much the same position; it is in almost total chaos as people search for a new form of harmony.

Would it be naive to ask whether there is a progression or just a change?

I don't know. I think it's primarily a change, but it's hard to say in what sense; there's a progress in some senses, but not in others. To define progress you must define a direction. If you choose a direction, you may discover progress, but if somebody chooses another direction, he may discover no progress.

Progress implies a target, an end-point.

Yes. I think there is no end-point and no target. The universe is an unending transformation in flux. Out of this appear these forms which have subsistence. The hardest thing of all is to see that we ourselves are only a form in this. The major reason, I think, why people find it difficult to accept this view is that it implies that we ourselves are only transient forms. The thought of the self has always been built around the idea that the self is an eternal substance, either material or spiritual, or both, and sometimes called the soul. I think that our views of matter and our views of ourselves are implicitly related. If a person is reluctant to believe that he is not a substance, he will be reluctant to believe that matter is not a substance.

This brings to mind what many mystics in the Middle Ages said about the spirit returning, rather than being permanent. Wasn't it Meister Eckhart who believed that God was a Negative or a Nothing, and one returned to this?

Many religions have had that view, or something like it. In older times, people put their philosophical views in religious terms. The separation between art, science, and religion was a more modern development. Consider these ancient religious interests: one was the origin of things, the general structure; every religion had an explanation of this structure. First men see the world; then they see themselves as separate from nature in the world; then they somehow conceive a unity between the two in the process that created both nature and the world. This is the canonical form which people must come to. Then they invent various myths as to how this came about; they become attached to those myths; and the myths are overthrown.

Scientists have invented other, shall we say, myths as to how it came about, for example the astrophysical story. Eventually this will fade out into something unknown too. People are always trying to understand this wholeness which they seem to be separated from and trying to explain it. The way I think of it is this: suppose we take it hypothetically that when man was just coming from the animal stage, he never thought of himself as a separate being, as separated from nature in any way. At some stage man began to think 'I am myself separate, I am a substance.' That may have given him some positive advantages but, at the same time, it gave him the negative feeling that he was separate from everything else, and he felt weak, lost, and alone. Therefore, man began to search to unite himself again with that from which he thought he had separated. In doing that, he invented various mythologies as to how man and the universe were created from some common source. Man is still pursuing this, but in scientific terms, rather than mythological. You notice that astrophysics gets tremendous support, not only because it's interesting, but also because it touches on this. It means a lot to these people that they are explaining their own origin in common with the origin of the universe. That gives them a tremendous impetus to do the work.

This reminds me of the celebrations in Washington (1973) for the 500th anniversary of the birth of Copernicus [National Academy of Sciences, now published]. *It was an almost mystical religious meeting in the end, the way that the scientists talked about Copernicus, as if he were a saint or a god.*

One root of the word *religion* means 'to bind up, to unite'; the other is 'holy,' which has the same root as *whole*. Man feels separated from everything and he is always trying to bind himself back to it, to make it whole again. An ancient philosopher said that man's activities could be divided into three basic kinds: the scientific, the artistic, and the religious; science dealing with knowledge, art dealing with harmony or fitting, and religion dealing with this search for oneness. Scientists are still searching for oneness and so are artists in some way. Religion has become highly fragmented, and people no longer believe in mythology. The fact that people with religious intentions looked into these questions in the past is not at all surprising. In fact, there is no separation. I don't think that you will ever get rid of this search for unity, which was one side of what men meant by religion.

You mentioned the fragmentation of thought which has taken place over the last few centuries.

Thought has a tendency to fragment, to look at the world in little pieces. The situation is very extreme now, with so many different subjects of study in the universities, and none of them connected. Some people try to make interdisciplinary subjects, which in turn become more fragments. I think the general fragmentation of knowlege is producing a problem today. Once there was the idea of the whole of knowledge, but that's obviously vanished long ago. Thought has an inherent tendency to produce fragments, to focus on one thing and then on another, then another. That is even necesssary for good thought.

Could you connect this up with the idea of implicate and explicate order?

Various fragments are explicated by thought. In a hologram, you could fold up a tremendous number of pictures and any one could come out. That would be a fragment, and it would look like a whole, but it wouldn't be.

Is it true to say that conscious thought is explicate thought?

Yes, it is fragments being made explicate. You could say the unconscious is this vast background, which is ambiguous and cannot be defined.

Does consciousness necessarily imply fragmentation?

It depends on what you mean. The content of our thoughts involves many fragments. I think that is inevitable. We have got to focus on this problem or that, and we must separate one thing from another. We cannot try to do everything all at once. But our thought is not merely an image of things, it is also, more deeply, a disposition to act in a certain way. If we have fragments disposing us to act in different ways, that will start tearing us to pieces. We can see this happening in society, where all sorts of different views exist, and people are going in all sorts of different directions that are not compatible. Conflict arises either within one person or between people. At this point, people wish to establish wholeness, and they may try to impose it through some philosophy or some religion or some political theory as the order which will establish wholeness. It is actually only another fragment. What we want is to have wholeness in the activity of the human being, while the thought can fragment as much as it needs to, to deal with each particular aspect. At the same time, there will be all the different fragmentary views, and we must try to develop some broader views, not to impose them and say they are truth, but to see things more broadly, at the same time that we see them narrowly.

Historically we would probably say that this fragmented way of thinking was an evolutionary process, as man confronted nature and tried to survive. Do you suggest a new evolutionary step in thought?

I don't say exactly that. Man had certain survival advantages by breaking things up, fragmenting them, treating them separately. But there are also disadvantages, as people are discovering. When you treat nature as fragmentary, dealing with one fragment after another, various problems occur, such as pollution or exhaustion of resources. It is not clear that fragmentation is an unalloyed means of survival. But this does not imply a simple return to the time before man knew his separation from nature. Once man has had the thought that he is different from nature, he can never return. There is an inherent contradiction in the assumption that man can return to nature, which makes it impossible. If he tries, he will start with his mind, which is supposed to be separate and struggling to unite. But the very struggle to unite will be an expression of the fact that he believes himself to be still separate. This is the contradiction. What man *can* do is to get *beyond* that thought. Until man had the thought that he was different from nature, there was no fundamental disharmony. But the disharmony arose when man thought that he was different, isolated in some sense, and therefore in need of reuniting. If you think it over, what he is trying to do is to reunite what has *never* been separated.

There has to be a very big change in our way of thinking. I believe that quantum mechanics and relativity both point, to some extent, to what step is needed. Fundamentally, the step is to be free of this division between the self and the world, the observer and the observed. I think all our thinking tends to be based on the idea that thinking is carried out by an entity, who could be called a thinker, a self, or an 'I.' As Descartes said, 'I think, therefore I am.' He was only expressing what people had felt for a long time. He did not invent that. One view is to say that thinking is carried out by a mental or spiritual entity somewhere inside the body – the 'thinker'; that the thinker produces his thoughts, but is separate from this thoughts. Since the thinker has clear differences and properties from, say, the table, you must say that the thinker is a different kind of substance than ordinary material substance. That is what Descartes said. There are two kinds of substance: one is extended substance, ordinary matter; the other is thinking substance, which is mind. Once you have introduced this idea of separation and fragmentation, you must inevitably come to fragmenting the thinker from his thoughts, and from the world that he is thinking about.

But that separation is false and illusory, and the notion that there is a thinker inside who is producing the thought is merely imagination. What would be closer to the point would be to say that there is nothing but thought, and no 'thinker' to produce it.

So this whole process of fragmentation is a process out of nature?

But is it an actual process? Is it not an illusory process?

But there is pollution, exhaustion of resources, and other problems.

Yes, but it is an illusion that there is any fragmentation in a fundamental sense. These problems are actually an expression of our *oneness* with nature, not of our difference. Man's thinking tried to be different from nature and approached it in a fragmentary way, trying to treat it as pieces. But nature refuses to be treated as pieces. Man thinks these illusions, and his mind being disposed by the illusions, he creates real action which is out of harmony with reality.

So this fantasy, this illusion, is turning into a nightmare?

Yes, the fantasy produces real activities which are destructive. It is because man is one with nature that this happens.

Yes, but all these activities are part of the processes of nature.

Man's thought is part of nature, and when man's thought goes into fantasy and mistakes it for reality, that is also part of nature. The trouble is that man has the illusion that reality as a whole is fragmented, instead of seeing that it is his thought which is fragmented. Thought is like a bunch of maps. The maps are fragments, but you don't imagine that the world is fragmented because the maps are. It is useful to fragment these maps because it enables you to focus on details.

If pollution and so on are part of the illusion, then what you mean by reality is very subtle.

Reality cannot be specified unambiguously. It is the flowing, an eternal transformation. Transformation cannot be pinned down unambiguously. It is movement, which means that any attempt to pin it down is an illusion. Thought makes these fragments which do appear to pin it down, and they are useful, but it is an illusion to suppose that the country doesn't change because the map doesn't. If you have a map published fifty years ago, and try to direct yourself through London, you will have trouble. These maps are fragmentary not only because they are broken up into pieces, but also because they are based upon the past, and the past is a fragment. The fifty-year-old map of London can give you only a fragmentary picture of the situation now. If you supposed that the situation never changed, then it would work.

Can our thoughts escape once they are tied to language?

Yes, they can, because we can look into the language. A language not only expresses our thoughts, but also helps work back on those thoughts, and gives them some fixity of shape.

But if our language is inherently fragmented, it becomes increasingly difficult to free our thoughts.

I think that our thought is in fragments in the first place, and that is why our language is fragmented. I don't think that the trouble can be described as originating in language. It originates in the very nature of thought. Language can be considered only as a secondary process. We have developed the language which emphasizes fragmentation by having one word for an object and saying that the object acts on another one and so on.

Is it possible, then, to understand quantum mechanics or the world within the language we use at present?

I think we can, although we might also change it. Language is always used figuratively and poetically, I think; we never use it literally. The attempt to give unambiguous significance to language will never work. It is inherently ambiguous, it is flowing, the meanings are flowing. If we think differently, we will find ourselves using the words differently. Perhaps, ultimately, we will change the formal structure as well.

You have worked with language structures yourself.

I made some experiments trying to change the structure of the language, just to see what would happen. I emphasized verbs instead of nouns, to emphasize the flowing movement. I saw that you could actually do quite a bit on that line, but I finally felt that you couldn't push it too far if you made a special language, because you would merely create another fragment. Special languages have been made and they have had a fragmentary effect.

I remember thinking that it seemed to be a language very constrained by sets of rules that you were developing.

It was more a mathematical kind of language. I was trying to 'mathemate' the language so that we would not have such a sharp separation between mathematics and ordinary language.

I never quite understood what your reservations on relativity were.

They are related to what I said previously about time and movement. First of all, relativity takes the space-time continuum for granted, which implies that time is a substance, or something you move through. I don't think that makes sense, you see. There are many ways in which relativity and quantum theory need to be changed together. There are two very elementary points which quantum theory does not deal with. One is the existence of things. Quantum theory says that nothing can be discussed except the probability of what will be observed when you have a piece of equipment. If we take the whole universe, we would have to suppose another universe of observing equipment, perhaps bigger than the first.

Nevertheless, we must say that, in some sense, the universe does not require that universe of equipment; it is there without it. As Bohr said, classical mechanics did not explain that atoms are there, the most fundamental thing, the clue to something new. Quantum theory does not explain that matter is there without a tremendous amount of equipment to specify its state.

I think that every structure abstracts some things which are really folded up in the totality. They have some relative subsistence, but the attempt to say that it covers everything is going to make it impossible to be consistent.

The second thing that quantum mechanics does not discuss is the *actual process*. For example, if you take a single radium atom decaying into a Geiger counter, quantum mechanics proposes a wave function, half of which leaks out of the atom in two thousand years. However, in some cases something happens immediately. Let's say that it takes ten years to decay. Since the counter doesn't work for the first ten years, you know that nothing has happened, and that the wave function is entirely inside for the first ten or one hundred years, or whatever. And this contradicts the idea of Schrödinger's equation, which says that it was leaking out all the time.

The theory says that Schrödinger's equation is the most complete description possible. I say that must be wrong, and that Schrödinger's equation is an abstraction of a fragment.

At one time you tried to look at it using the notion of hidden variables.

That is just one way of saying that there is more to it. That was perhaps too classical an approach. It was merely a way of getting insight.

Was there not some misinterpretation of what you were trying to do?

Yes. I think that some people thought that I was trying to return to classical concepts, but I was really using the hidden variables to get imaginative insight into what the theory meant. One could see that the hidden variables would have certain peculiar properties which suggested that you should look at it in another way. For example, one discovers that these hidden variables have properties which imply instantaneous connection of all parts of the universe, an extreme form of wholeness. Some people have said that it is so strange that they do not want to consider it. I don't think that it is sensible to say that as long as we do the computation, we don't have to imagine anything about it, so we are not disturbed by anything that happened. I don't understand the attitude which says that hidden variables have strange properties and therefore we would rather not use them. By using hidden variables, your attention is focused on these

strange properties, and by understanding quantum mechanics imaginatively, even if as not yet fundamentally, you begin to see that quantum
mechanics implies something very new, which you are missing by just
doing the computation.

*At this point a traditional sort of physicist would ask you to produce a new
prediction.*

That again is a sign of a certain attitude to physics, which says that the
essential point about physics is to predict something. Why do you want to
predict? You would think that there is a predictive instinct which must be
satisfied. But this is obviously not the case. The reason why people want to
predict is just to confirm that their ideas are on the right track. I am trying to
say that in some cases you cannot predict; some things are ambiguous.

Trying to see the weakness of a theory in a traditional way is inadequate. The traditional scientific method is to say: wait until your experiments clearly show that you are wrong. But if you are going along with
confused methods, no experiment will clearly show that you are wrong,
because you can always modify your theory. This has often been done.

We must look at it differently, realizing that there is something wrong,
which the present theory does not have in it, which requires understanding, namely, there is an actual individual event – the decay of the radio-
active nucleus – which is simply not accounted for in the present theory.
We must put in new concepts to account for it, and see what happens,
even if we can't use them to predict anything more at the moment. I think
that there is an overemphasis on prediction, on getting results, which is
stifling physics. Many people don't fully and deeply realize that there *is*
something missing. They are so used to doing statistical calculations, and
saying that only statistics matters, that they do not notice that there is an
actual, individual fact which is not accounted for.

Suppose we begin by saying that stationary states just simply exist, and
that the world is very nearly in a stationary state, with some transitions.
Now these stationary states make jumps from one to another. We haven't
explained why that happens, we are only accounting for that fact, in which
case we do not need equipment at all. We are not going to interpret quantum mechanics as what an observer would see, but as a process of jump-
ing between quasi-stationary states. We now explain that we have what we
call a material system, which is a stationary state of a large number of
atoms. When you solve the many-body wave equation of quantum
mechanics, you see that you cannot make this relativistic, you need to
have one common time for the whole system. However, this material
system is essential for relativity, because the theory presupposes some
quasi-rigid material system as a frame from which to make observations. I

don't think that you could ever get any definite meaning to relativity for a single atom. You would not have a clear definition of properties such as direction in time for example.

I believe that Roger Penrose was trying to take collections of atoms, and, in that way, he thought that he could define the direction.

Penrose is working on some particular mathematical structure to try to do something which may well be worth doing. I am trying to discuss something else. I think something has to happen at a lower level as it were. Physics has given us some facts, but the actual language of discussing these facts is confused.

I am saying we must consider that this piece of apparatus, this block of matter, exists, without any help, as it were, of observers, or anything. It is in a nearly stationary state, and it determines a frame. That is missing from relativity theory; there is no clear definition of a frame. If it were not for quantum mechanics, which makes matter stable, there would be nothing in relativity that would allow for the frame that measures anything. So there seems to be a deprivation of relativity in quantum theory.

To pursue this further, you find that, if you take stationary states in one frame and then move the system and accelerate it, the stationary states of that system are not compatible with those of the first. They are nonstationary, and they correspond to operators which are not compatible. So if you have one system in a stationary state, and another system moving, the moving system is *not* in a stationary state relative to the first system, but it *is* relative to its own frame. If there were an observer inside, he would be built out of atoms in those stationary states, and he would see everything relative to those stationary states, that is, as changing. So we have an interesting conception: a relativity of stationary states. What is stationary for one block of matter is not stationary for another. That concept has been missing. In fact, we have to give the same 'time' to all the atoms in one system, since we say there actually is a common time which is relevant for determining the stationary states of the first system, in technical terms, as the time-displacement operator for that system. Another system has another time-displacement operator and determines another set of states. The two systems are not stationary together.

Now, we get an extension of relativity, because we have introduced a new concept, which I call the material frame, or the natural frame for the stationary state. However, this means that Schrödinger's equation cannot be taken as a complete account any more, because each system has its own Schrödinger's equation, giving its own stationary states. One should not try to say that a single Schrödinger equation covers the whole universe.

We say that a particular material frame, or block of matter for example, is determined by solving Schrödinger's equation for some atoms. This can be extended abstractly to the surrounding space, and another set of atoms which has its own frame. They may be related approximately but not exactly.

Then is space a relational notion?

Space comes out as relational space. Every particular block of matter has its space, which is extended abstractly into the surrounding region, and all these different spaces interpenetrate to form what I call a *multiplex*, that is, many spaces folded together. Rather than thinking of space as a single substance, we can think of it as an abstract relation of the multiplex, each element of it being a fragment that is based on one piece of matter. That is an order we are imposing on space and relating to matter.

These spaces are not all the same. We can approximately replace them by one bigger space, which some have done, and call this a real space. But I say that it is no more real than the smaller space, although it may be convenient for some purposes.

The total implications, or the metaphysics of quantum mechanics and relativity together, are utterly unclear. Every fact, you see, is presented in a framework of a set of concepts and ways of thinking. If you have a confused order of thinking, the fact will be confused. So I'm saying that the first step is to get a clear presentation of this fact. Then we can go on to develop mathematical methods of going further. When we have a set of facts, our next step is to develop a broader mathematical way of thinking, which will assimilate those facts as aspects of the mathematical concepts. I don't think the present fact is clear enough to assimilate into any mathematical system. That is one reason why so little progress has been made over the past forty years.

Carl Friedrich von Weizsäcker

Professor Weizsäcker (1912–) is Director of the Max-Plank Institute at Starnberg, Germany. He began academic life as a physicist and his abilities were soon recognized by Werner Heisenberg, who co-opted him to his team of brilliant young scientists. As a member of this research group Weizsäcker was to make important contributions to the theory of nuclear structure.

Professor Weizsäcker, like his mentor Heisenberg, has a deep interest in philosophy which has taken him to the Chair of Philosophy at Hamburg University. To the professions of philosopher and physicist can be added a third: political scientist.

We spoke with Professor Weizsäcker at his institute, located beside Lake Starnberg in the mountains near Munich. Its tranquil setting seemed appropriate for a philosopher and physicist.

DP *Professor Weizsäcker, you begin a discussion of quantum theory with a logic which emphasizes the distinction between past and future. Would you elucidate this emphasis?*

Perhaps the right way of answering your question is for me to say two different things about the way in which I was induced to work in this field, first of all, long ago, in connection with thermodynamics and later on in connection with quantum theory. Now, when I studied thermodynamics in the university, I found it very difficult to understand how the irreversibility of actual events can be reconciled with the reversibility of events according to the basic laws of mechanics. I was informed by my teachers, and by the textbook, that this was done by introducing the concept of probability, of statistics. And then I wondered how the concept of probability could introduce an asymmetry into time if it wasn't there from the beginning. The answer which I think I understood in the end, and which I think is correct, is that the asymmetry is brought into statistical thermodynamics by the fact that a probability of an event, in the direct sense, is always the probability of a *future* event. You ask 'how probable is it that it will be raining tomorrow?'; you don't ask 'how probable is it that it will be raining yesterday?' You cannot even express that in meaningful English. You can ask 'how probable is it that it was raining yesterday?' but that has a completely different meaning; it means that you do not know whether it was raining or not, and you want to know how probable it is that you *will find out* that it was raining. So, again, it refers to the future.

DP *Are there not events in the past which are no longer testable, but about which we can make statements in terms of probabilities?*

Even there, I would say if they are actually not at all testable, it is meaningless to apply probabilities to them. If you are not able to test it now, and you say there is a probability of five per cent that it happened like that, and then by some good chance you find a way of testing it, then your probability applies. But I would say that I would be prepared to defend the view, in a discussion which would last two or three hours, that probabilities basically always refer to the future and all other uses of the term probability are made in a more or less oblique sense. The probability of a future event, in my analysis, would be the most primitive sense of probability.

Then, the next step was that I also had some difficulties in understanding the basic ideas of quantum theory. Of course, I could easily understand the mathematical formalism of it, but what it *really meant* – that was always the difficulty. I found again that the concept of probability was used, so I tried out the hypothesis that this would again mean that it actually refers to the future. And I think you can say so, because if you speak of the probability of finding an electron at a certain position, it is certainly the probability that you *will* find it there. I found that this gave a possibility of perhaps understanding the fact that the probability calculus of quantum theory actually differs from the probability calculus which we learn at school or in university, which rests on, for instance, Kolmogoroff's axioms. This is a fairly technical point, and in general, I think, our students are not told about it. The fact is that the axioms of probability, as we learn them in ordinary probability courses, are not in agreement with their use in quantum theory. In quantum theory we have the superposition of probability amplitudes, of which nothing is known in classical probability calculus.

DP *So events in the future are described by this temporal logic, while events in the past have testable propositions.*

You can say that for the *past* we have testable propositions, we have facts, and facts are known or unknown. (Nobody would deny that a fact that is *un*known is a fact anyhow.) But for the *future* it would not be permissible to say that a future event, which is unknown, is an event anyhow. And this is precisely the point made by the so-called indeterminism of quantum theory: that if you assume that future events are objective, even as long as they are in the future, then you would have to presuppose the classical probability calculus, Kolmogoroff's axioms, the Boolean lattice of events, etc., which is not, in fact, the structure which quantum theory actually has. But if you say that a future event is actually not an objective event as long as it has not happened, but is a possible event (and possibility means something which is not just 'actuality which is not known'), then you avoid every contradiction and you can interpret quantum theory in a common-sense way.

DP *Are these future events the potentialities that Heisenberg talks about?*

Yes, I would say they are more or less the same thing.

PB *Are they potentialities rather than possibilities? I mean by possible events those events which do not necessarily have any direct causalities, or even if they do there is not necessarily anything implicit in the present which links them directly. I am thinking of a seed that grows into a tree, so I would say that the tree is a potentiality rather than a possibility.*

If you offer this distinction, first I would ask you: would you take potentiality to be a special case of possibility? Would you say all potentialities are possibilities, but not all possibilities are potentialities?

PB *Yes, I think I would.*

In this case, I would say that the general theory of probabilities is a theory which refers to possibilities. Probability is a quantification of possibility, but where you have laws of nature which make it possible to predict probabilities from a given situation, this means that you have precisely that connection between present and future which you have been describing by saying potentiality. And in this sense, I would say here the word *potentiality* really applies.

DP *You have made a distinction between past and future, yet you do not appear to have introduced time, in the ordinary sense, into your logic.*

I have not introduced a metric of time, I have introduced an ordering in time, and this is a point which I think is quite important. To come to a more basic level of discussion, I feel that there is no possible use of concepts at all in the empirical world; there is not even a meaning to the word *experience* which would not presuppose the distinction between past and future. If you speak about experience and say 'I learned that from experience,' this means from past experience, of course. Or you say 'this is a very experienced person,' and that means he has learned things in the *past* which he is able to apply in the future. But you can never apply anything you have learned in the past in the past! That is meaningless. And you cannot have learned something from the future, except perhaps if you are a prophet. So I would say that the concepts which we use in common sense language when we speak about experience always presuppose the difference between past and future. Therefore I feel that whatever else we are going to introduce (like measurements, metric, space, objects) we always presuppose the difference between past and future. We presuppose it in such a manner that, in general, we are not even aware of our presupposing it. I call *this structure* by the name of time.

DP *This theory seems to involve a non-spatial aspect of time.*

Yes, I would say so. If you try to build up a consistent axiomatic structure in physics, and begin with the quantum theory, I have not found it possible to write down axioms for quantum theory without using time in the sense in which I have used it now. But you can very easily build up quantum theory without *any* use of space. An axiomatics of quantum theory can be done in this way. You only speak about measurements and probabilities, predictions of probabilities, and laws for such predictions. Then you write down a correct theory of quantum probabilities which would include the superposition principle. It is never necessary to specify what sort of measurements they are and whether these measurements are made in space or in some other manifold of possible states. Of course we know empirically that measurements are always made in space, and this must come out in the end, but the axiomatic structure of quantum theory can easily be made in such a manner that space is introduced at the very end.

I would say that time has a priority over space, but this priority is probably only possible as long as you speak in terms of quantum theory *without* relativity. The step which has not yet been taken in fundamental physics is not just to introduce a formalism which treats space and time on an equal footing, but also to understand how this happens.

DP *It seems that the tense logic you are developing is more general than the propositional calculus.*

First of all, you can say it is introduced as a variety of the propositional calculus. As soon as we go into the real mathematical problems – the problem of mathematical logic – then I would say we come to a problem which can first be formulated by repeating an objection. The objection is this: von Neumann, you, and others are talking about quantum logic and calling it a tense logic. But this is nonsense, because it is derived from quantum theory, and quantum theory has been built up by means of classical logic. So, how can you get a result which implies a non-classical logic by means of classical logic? This seems somehow to be a vicious circle, or a vicious non-circle! I take this very seriously! I would not say that if the description given by my adversary were correct, I would still uphold the idea that there is a quantum logic; *rather* I would say this description is not correct in the following sense: quantum theory has not been built up by a consistent use of classical logic, but by a fairly inconsistent use of classical logic. This inconsistency has been discovered, and has been corrected by saying that, if we want to do it correctly, we must do it with quantum logic, with temporal logic, tense logic. But this means that quantum theory cannot be taken to be just the mathematical formalism, but the formalism *with its semantics*, and with its meaning, because the formalism in itself can certainly be described by classical logic as far as any mathematics can be described by classical logic. If this is so, the question is whether tense logic, temporal logic, is a special case of general logic, or whether general logic will have to be explained by, or be founded on, temporal logic. My proposal is that the latter is the case.

In this respect, of course, I am following the intuitionists. I say that if you wish to understand the fundamental problems of mathematics, you will have to decide how to treat infinite sets, and my personal predilection here is operationalism or intuitionism, saying that the actual meaning of infinite sets is only the possibility of having certain constructions. If I take this view of mathematics, I apply the concept of time to the foundations of mathematics because operations are done in time. So I would flatly deny the non-temporal nature of mathematics.

DP *Time has been introduced into the theory in a very fundamental way. Could you explain how space is to be brought into quantum theory?*

The question of introducing space is precisely an element in the theory which really has yet to be achieved. I am now entering a field in which I am offering my own hypothesis in physics. This hypothesis is that space is connected with metrical time; that measuring time is closely connected to measuring space, and that this is a different level in the construction of physics than the level at which we have just spoken about past and future.

I think the mathematical nature of what we call space in physics can be deduced from the quantum theory of what I like to call the simple alternative: the 'yes-no' decision. The quantum theoretical description of a simple alternative is any experiment to which there are just two possible answers. Either the particular particle is a proton or a neutron. Or, there are two holes in a screen, and the particle which has gone through the screen has gone through either hole number one or hole number two. The quantum theory of simple alternatives is described by a two-dimensional complex vector space. This complex vector space is isomorphic, up to one sign, to three-dimensional real space. My hypothesis is that what we call space in physics, the space of possible positions of particles or fields, is deducible from quantum theory by being identical with Euclidean (or perhaps non-Euclidean) three-dimensional space, which corresponds to the two-dimensional vector space of the quantum theory of the simple alternative.

DP *The idea of a space built out of simple alternatives, binary logic, seems similar to Roger Penrose's attempt to derive space from a matrix of spinors which themselves have binary values.*

I would say that it is very close to Penrose, and it seems to be independent. I always like to discover somebody else who did the same thing. The probability is a little bit greater, then, that it might be true.

DP *So in the end you have a tense logic and a three-dimensional space. Do you feel it is necessary to go to the four-dimensional space-time of general relativity?*

You cannot say I have a three-dimensional space yet. First of all, I have quantum theory, including the concept of time in the sense of tenses, and I have two-dimensional complex vector spaces, and these can be somehow reduced to three-dimensional space.

DP *Do you have a continuous metrical time?*

If I accept quantum theory in the way in which it stands at present, I have a metrical time. I have not *justified* it, but I have it. And this is just one of the points which I try to clarify. How can I defend, in the end, the semantic consistency of speaking of metrical time which I did from the outset in building up quantum theory and an axiomatic system? I would propose that the parameter time, which we use in quantum theory, is only defined as a classical limiting case; it is not an observable; as we all know, we cannot describe it as an operator. If we speak about measuring time, and we describe that by real quantum theoretical measurements, probably the corresponding operator is not identical with the parameter time. The parameter time is always external to the system.

DP *In the description of experiments involving very small distances, would your theory give results different from conventional quantum theory?*

I would be very happy if I could answer your question and I hope to answer it in the end, but at present I would just say this: if I introduce space in the way I did it now, the real problem is not yet solved. This is just the first step, the step in which I show that in addition to the parameter time I can also introduce a parameter space.

But then I must speak about measuring time and measuring space. I would have to describe space as something (if I may use slightly picturesque language) which originates in the interaction of those physical objects which we then call particles. I would not say that there is space which can be subdivided indefinitely. In the parameter space, you can describe it like that of course, but that means that you have to produce more and more particles. As long as I have a finite number of particles, I have a limitation to the possible subdivision of space. For instance, if I say I have ten to the eightieth particles in the world, it would not be possible to define a smaller length empirically than ten to the minus ninety-three centimetres, because you would have to use all the particles in order to measure that.

PB *What does the word* fundamental *mean to you? Can we talk about a fundamental level?*

Fundamental is probably always a relative term; something is more fundamental than something else. That elementary particles are *not* fundamental is well known today, because they are changed into each other. I think that if we do not try to go beyond the frame of quantum theory, then the most fundamental physical objects would be objects which admit of only two possible answers to a question. I think that you cannot subdivide quantum theoretical systems or objects beyond that. These most fundamental objects which are admissible within the frame of quantum theory would be the things we must study. In this sense, within the frame of quantum theory, I cannot think of anything more fundamental than a simple alternative. And this alternative is in the sense of being fundamental far beyond space. Space has to be built up from such alternatives, and not the other way around.

PB *Why has it taken so long historically for man, who is totally enmeshed in time, to be brave enough to include time right in the very foundations of mathematics?*

That is a most interesting question! Like all interesting questions, it probably has not precisely one answer; one must say several things. I would have two proposals, which are not contradictory but complementary to each other. One is, and this is true also in philosophy, that the simplest things are not conscious, are not in the conscious mind, but are presuppositions of which we are not even aware.

It is just the nature of time that whenever you try to say anything, what you have been doing is already presupposing some understanding of time. St Augustine said: 'if you do not ask me about time I think I know what it is, but if you ask me, I am not able to answer.' This is, I think, closely connected with the fundamental role played by time.

If you assume, as I do, that the concepts describing time belong to the most basic concepts, then it is not probable that it will be possible to explain them by reducing them to anything else, that is, to any other concepts. Rather, the other way round. Then one would expect that it would need a thoroughgoing analysis to come to a description of the structure of those most fundamental concepts which I suppose are closely connected with the concept of time.

This is only the one answer. The other answer is this. Certainly everybody has a good understanding of time, not perhaps to say *what it is*, but to use all the concepts correctly. Tenses, in Indo-European languages, are used quite correctly even by children, although the greatest logicians have great difficulty in explaining what the children are saying. There was, however, a particular step taken in Greek philosophy: the attempt to eliminate time from a fundamental role in it, and to replace it by concepts which are *beyond* time. This philosophy, which was closely connected, I think, with mathematics, was so successful that it influenced all later thought in a manner which has detracted awareness from the temporal relations which are involved in it. This is just one of the great steps in philosophy: that people learned to explain the world in a manner which tried to reduce it to something beyond time.

PB *Human beings feel time very deeply, yet the Greek ideal was to explain ourselves and the universe in non-temporal terms. This implies a perspective that isn't human: it does not contain man implicitly. It leads to equations and concepts which do not have a human face, if I may use a poetic image. But the quantum principle makes it more and more difficult to eliminate ourselves from the universe; and the introduction of temporal logic, or time, into the very foundations of mathematics seems to be bringing, in a strange way, human beings back into the universe.*

Yes, I think that is a very good description. You could say that the Greek attempt is an attempt at divine knowledge, divine understanding, not *human* understanding. But that is not the whole truth, because, if you understand Plato or Aristotle, you find that they were fully aware of their own human nature, their limited understanding, and they were also aware of the very profound role of time. But time, as far as it enters that philosophy, I think, is always understood in the image of a circle, its returning into itself. The highest form of motion is circular motion, that is, a motion which never leaves itself. In this sense, it is an interpretation of time, but a special interpretation – an interpretation which does not take account of irreversibility, of what we call the second law, or of what we call evolution. That is one point.

Another point is that if there is a self-contradiction in classical metaphysics, which I think is absent from Plato but is very much present in much of later metaphysics, it is that these people think that, as human beings, they are able to see things with a divine eye and then able to formulate concepts which can be defended 'on the market' so to speak. I think this is somehow bringing the divine world-view too much into the human sphere, and forgetting our human limitations. In modern times both Hume and Kant, very different thinkers in very different traditions, have formulated quite clearly that our own theory of understanding must be a theory of a limited, finite understanding. Quantum theory is precisely the step in physics which makes it no longer admissible to forget about the human nature of the one who is making the theory – not only the observer, but the theorist. It might be that for a truly divine understanding it would still be true that there is a basic reality beyond time. But this cannot be formulated in all the nice little concepts of logic and mathematics which we are using. These concepts belong to human beings, and they belong to time.

Paul Adrien Maurice Dirac

In one of C.P. Snow's early novels a character in the scientific life of Cambridge is described as the successor of Newton. It can only be Paul Dirac (1902–). Like Newton before him Dirac has made contributions that are respected by his colleagues not only for their depth of insight and clarity but for the power and economy with which mathematics is brought to bear upon the problems of nature. Dirac's scientific papers have the polished and balanced appearance of a sculpture by Brancusi.

While Heisenberg was discovering the principles of quantum mechanics in his Helgoland retreat, Erwin Schrödinger followed a different path to derive his wave mechanics of the atom. Dirac was able to show that the two theories were equivalent, and in the process provided quantum theory with a sound mathematical footing. His contributions in physics also include the quantum theory of matter and radiation, the prediction of the spin of the electron, and the existence of the positron as well as an attempt to form a marriage between quantum theory and the theory of relativity. He was awarded the Nobel Prize in 1933.

Professor Dirac has retired from his position as Lucasian Professor of Physics, the chair previously held by Isaac Newton, and is at present at the Institute of Advanced Studies at Miami, where he was interviewed. At first Professor Dirac seemed reticent about his achievements in physics. However, when the topic of beauty in physical theories was raised, Dirac began to speak with animation. In addition to commenting on the current status of theories of physics he touches on the 'large number hypothesis,' which has occupied him in recent years. Dirac is concerned with the occurrence of large numerical constants in physical theories. Rather than ignore these numbers or ascribe the similarity of their values to mere coincidence Dirac has proposed that constants of nature are interrelated.

The large value of certain of these constants, Dirac supposes, is connected with the age of the universe.

DP *Do you feel that there is the same excitement today in physics that there was in the twenties and thirties?*

The problems are more difficult now and there is not the same hope of making rapid progress which there was in those days. Excitement is usually combined with the hope of making rapid progress, when any second-rate student can do really first-rate work. But the easier fundamental problems have by now all been worked out. Those that are left are very difficult to work on, and one doesn't seem able to get the right basic ideas for handling them.

It is quite possible that they will require wholly new ideas. In fact it's pretty certain they will; otherwise they would already have been thought up.

PB *But they will still be related to the existing development of theory in some sense at least.*

Yes. The present theory must be an approximation to any improved theory which we get in the future.

DP *Some people we've spoken to seem to think it's a matter for new experiments, particularly in elementary particle physics.*

If the theorists are not good enough to solve it on their own, that's what one has to do. It needs an Einstein, or someone like that. Einstein didn't depend on new experiments to get his ideas.

DP *Do you feel that the progress in particle physics is fruitful?*

It's not really fundamental; it's collecting a mass of information and one doesn't know really how to get the basic ideas from it. Just like in the early 1920s one had a mass of spectroscopic information and it needed Heisenberg to find the real basis of a new theory from that wealth of material.

DP *Do you think a unification necessarily will have to include relativity?*

I should think so, ultimately. Perhaps not gravitation in the first place; gravitation is rather separate from ordinary atomic physics and it plays very little role.

DP *It seems to be an insurmountable problem to most people: the quantization of relativity. It is something you have worked on.*

One can deal with it up to a certain point, but one cannot complete the theory in a satisfactory way.

DP *Could you summarize your thinking on the large numbers hypothesis?*

The large numbers hypothesis concerns certain dimensionless numbers. An example of a dimensionless number provided by nature is the ratio of the mass of the proton to the mass of the electron. There is another dimensionless number which connects Planck's constant and the electronic charge. This number is about 137, quite independent of the units. When a dimensionless number like that turns up, a physicist thinks there must be some reason for it. Why should it be, well, 137, and not 256 or something quite different. At present one cannot set up a satisfactory reason for it, but still people believe that with future developments a reason will be found.

Now, there is another dimensionless number which is of importance. If you have an electron and a proton, the electric force between them is inversely proportional to the square of the distance; the gravitational force is also inversely proportional to the square of the distance; the ratio of those two forces does not depend on the distance. The ratio gives you a dimensionless number. That number is extremely large, about ten to the power thirty-nine. Of course it doesn't depend on what units you're using. It's a number provided by nature and we should expect that a theory will some day provide a reason for it.

How could you possibly expect to get an explanation for such a large number? Well, you might connect it with another large number – the age of the universe. The universe has an age, because one observes that the spiral nebulae, the most distant objects in the sky, are all receding from us with a velocity proportional to their distance, and that means that at a certain time in the past, they were all extremely close to one another. The universe started quite small or perhaps even as a mathematical point, and there was a big explosion, and these objects were shot out. The ones that were shot out fastest are the ones that have gone the farthest from us. That explains the relationship (Hubble's relationship) that the velocity of recession is proportional to the distance, and from the connection between the velocity of recession and the distance we get the age when the universe started off.

It's called the big bang hypothesis. There is a definite age when the big bang occurred. The most recent observations give it to be about eighteen billion years ago.

Now, you might use some atomic unit of time instead of years; years is quite artificial, depending on our solar system. Take an atomic unit of time, express the age of the universe in this atomic unit, and you again get a number of about ten to the thirty-nine, roughly the same as the previous number.

Now, you might say, this is a remarkable coincidence. But it is rather hard to believe that. One feels that there must be some connection between these very large numbers, a connection which we cannot explain at present but which we shall be able to explain in the future when we have a better knowledge both of atomic theory and of cosmology.

Let us assume that these two numbers are connected. Now one of these numbers is not a constant. The age of the universe, of course, gets bigger and bigger as the universe gets older. So the other one must be increasing also in the same proportion. That means that the electric force compared with the gravitational force is not a constant, but is increasing proportionally to the age of the universe.

The most convenient way of describing this is to use atomic units, which make the electric force constant; then, referred to these atomic units, the gravitational force will be decreasing. The gravitational constant, usually denoted by G, when expressed in atomic units, is thus not a constant any more, but is decreasing inversely proportional to the age of the universe.

One would like to check this result by observation, but the effect is very small. However, one can hope that with observations that will be made within the next few years, it will be possible to check whether G is really varying or not. If it is varying, then we have the problem of fitting this varying G with our previous ideas of relativity. The ordinary Einstein theory demands that G shall be a constant. We thus have to modify it in some way. We don't want to abandon it altogether because it is so successful.

I have proposed a way of modifying it which refers to two standards of length, one standard of length which is used in the Einstein equations, and another which is determined by observations with atomic apparatus. I should say that the idea of two standards of length and of G varying with time is not original. This sort of idea was first proposed by E.A. Milne about forty years ago. But he used different arguments from mine. His equations are in some respects similar to mine; in other respects there are differences. So this theory of mine is essentially a different theory from Milne's, although based on some ideas which were first introduced by Milne. One should give Milne the credit for having the insight of thinking that perhaps the gravitational constant is not really constant at all. Nobody else had questioned that previously.

DP *This theory has an important consequence for the creation of matter.*

Yes, the amount of particles – elementary particles, protons, and neutrons – in the universe is about ten to the seventy-eight, the square of the age of the universe. It seems again one should say that this is not a coincidence. There is some reason behind it, and therefore the number of particles in the universe will be increasing proportionally to the square of the age of the universe. Thus new matter must be continually created.

There was previously a theory of continuous creation of matter called the steady state cosmology, but this theory of mine is different because the steady state cosmology demands that G shall be a constant. Everything then has to be steady, and in particular G has to keep a steady value. Now, I want to have G varying, and I also want to have continuous creation. It's possible to combine those two ideas and I've worked out some equations on possible models of the universe incorporating them.

PB *One of the consequences of your theory is that it rules out an expanding-contracting universe.*

That is so, yes, because in the theory there will be a maximum size. This maximum size, expressed in atomic units, would give a large number which does not vary with the time. Now, I want all large numbers to be connected with the age of the universe so that they will all increase as the universe gets older. If you have a theory giving you a large number, of the order of ten to the thirty-nine, which is constant, you must rule out that theory.

PB *This implies a constantly expanding universe.*

Yes. It must go on expanding forever. It can't just turn around and contract, like many people believe.

PB *So that avoids the singularity at the end, so to speak.*

Yes, that is avoided; there is just a singularity at the beginning.

PB *There seems to be, or at least it's possible that one may observe such a thing as, a black hole, which is a theoretical consequence of general relativity. That is also a singularity, is it not?*

It depends on what mathematical variables you use. It would be a very local singularity anyway, not a cosmological one.

PB *But it seems staggering to the imagination that the mass of the star is concentrating into a smaller and smaller volume. I know there are repulsive forces that can stop it at various stages, but finally, I understand, with a star that is perhaps five or ten times the mass of our sun, it need not stop.*

That is what it seems, according to current theories.

PB *It is difficult to imagine such an object, but I suppose that is not a necessary condition for doing physics.*

If you can find equations for it, that's all the physicist really wants. It is quite likely that the laws will get modified under these extreme conditions; we'll have to try to find out what the correct laws are.

PB *But they need not contradict physical theory, wouldn't they simply be modifications?*

They would be modifications, modifications holding under extreme conditions.

DP *Would you comment on the divergences and infinities which occur in quantum field theory. Many think that they can be removed by renormalization. Is this your feeling?*

It's just a stop-gap procedure. There must be some fundamental change in our ideas, probably a change just as fundamental as the passage from Bohr's orbit theory to quantum mechanics. When you get a number turning out to be infinite which ought to be finite, you should admit that there is something wrong with your equations, and not hope that you can get a good theory just by doctoring up that number.

DP *Some people have suggested that by introducing curved space you can get rid of these infinities, Abdus Salam for example.*

I know that he is working on that idea, but I feel that with a good theory these infinities would never arise in the first place.

DP *The papers you produced have been universally considered beautiful. Were you guided by notions of beauty?*

Very much so. One can't just make random guesses. It's a question of finding things that fit together very well. You're solving a problem, it might be a crossword puzzle, and things don't fit, and you conclude you've made some mistakes. Suddenly you think of corrections and everything fits. You feel great satisfaction. The beauty of the equations provided by nature is much stronger than that. It gives one a strong emotional reaction.

DP *Do you get this reaction from certain branches of modern physics today?*

Not the renormalization theory, no!

PB *I have a question about the interpretation of equations. There are certain equations and certain theories where interpretations have been open to a great deal of discussion. It is not quite clear what's really meant in non-mathematical terms; I'm thinking of the principle of complementarity.*

Yes, there is an uncertainty in the interpretation. But I don't feel it is too profitable to discuss the uncertainty because the basic equations themselves are uncertain, as I was trying to explain to you previously. If you don't have very great confidence in the basic equations, then there's not really much point in spending a lot of time on the interpretation of the equations, as you believe they will be superseded after a while in any case.

PB *I was thinking of the uncertainty relations themselves. Do you believe that these will be superseded?*

It's possible. You'd probably have to pay a price for it and give up some other cherished idea.

PB *The problem of observation and measurement seems to be important.*

Yes, but you're discussing these problems on the basis of our present theories, which are just, I believe, a transient phase of physics and will be superseded after maybe a few decades – or, well, one just doesn't know when they will be superseded. It is rather as though one tried to build up a new philosophy on Bohr's orbit theory. You might have gone a long way with it, but all that argument would have been completely valueless when Bohr's orbit theory was superseded.

DP *If you were giving advice to young physicists today, which area would you suggest they look into?*

I think perhaps they ought to avoid fundamental physics because all the worthwhile problems there have already been very thoroughly explored.

DP *I mean in the sense of which area you think the breakthrough will come in?*

I don't know.

DP *You'd be there if you knew, I guess.*

Yes.

PB *Will it also depend on developments in mathematical theory?*

That's possible.

PB *In the 1920s the mathematics had to be partially invented as well, along with the experiments.*

The basic mathematical ideas were known previously to the mathematicians. They knew about Hilbert space; they knew about spinors. They had never thought that these things would ever have any physical application.

PB *So it's quite possible that some branches of mathematics already known contain useful approaches.*

Yes. However, an enormous volume of mathematics exists, and to look for which part is going to be useful in the future is pretty hopeless.

Roger Penrose

Roger Penrose's professional career began in pure mathematics. Born in Colchester, England, in 1931, he obtained a doctorate in mathematics from St John's College, Cambridge. His interests turned to the study of space-time structure and he spent a number of years at several American universities before returning to England.

His contributions in theoretical physics reflect his mathematical background for he seems quite at home when moving through multidimensional spaces, projecting infinities, or dissecting hyperspheres. With his imaginative approaches he has established important theories on singularities in space-time which have bearing on the nature of black holes. His insights into the structure of space-time are a valuable addition to the understanding of the theory of relativity.

Penrose was interviewed by David Peat in London, where he held a professorship at Birkbeck College; he has since moved to the Rouse Ball Chair of Mathematics at Oxford. His discussions in this chapter range from an appraisal of relativity theory to an explanation of his attempts to probe the nature of space-time using spinors and twistors. When he touches on beauty in mathematics it is clear that he is talking about something which is very concrete for him. Roger Penrose's hobby is the constructing ingenious mathematical puzzles and games. Those who have watched him in the process of constructing a mathematical theory realize that fun should never be the exclusive province of children.

Many people working on space-time structure accept the general theory of relativity or, at most, they are prepared to extend some aspect of the theory. Your own approach is more radical, for you are seeking a foundation for the properties of space and time.

I certainly think that one needs an explanation for the space and time that we see. In the first place one may reasonably ask to explain why it is we see just three space dimensions and just one time dimension. Many people would probably say that this is not a really meaningful question. I don't like to take that point of view myself. I think that it is a question that should be completely answered. In order to explain this kind of thing, one has to develop the idea of space and time out of more primitive ideas. When I say 'primitive' I don't necessarily mean that these will be ideas seeming more obvious to people; I mean they will be concepts more basic to the physics in some deep sense.

Perhaps I could mention one of the basic motivations for the whole thing. It is that one should ultimately try to get rid of the concept of continuum altogether in physics. There are really two basic places where the continuum comes into physical theory. The first and most obvious is in the structure of space and of time. There is apparently this continuum of space-time. The normal picture that one has is that between any two points one can find others and one can go on subdividing ad infinitum. No matter how small the region of space examined it essentially looks the same as it did before. When you think of it, this is really an absurd idea physically; you take a ruler twelve inches long and you cut it in half and you keep on doing this until you get down to the atoms and fundamental particles; so you try to slice them up. But we have no real reason to believe that space at that sort of level – if there is such a thing as a space at that level at all – is really like the space that we're familiar with. So I think that one really should question this use of continuum.

There is one other place where the continuum comes into physical theory in a really essential way, and that is in quantum mechanics. Here one has the complex continuum, where there are square roots of negative numbers in addition to real numbers.

Could you explain exactly how this comes into quantum mechanics?

I suppose the superposition principle is really the best way of expressing that. In quantum mechanics, if you have two states, you are supposed to be able to form other states which are physically admissible, by making combinations of these two states. So you can say, for example, if state A is allowable and state B is allowable, then state A plus state B is also allowable. The thing is, in quantum mechanics, that you also have to allow *complex* combinations. You have to allow state A plus the square root of minus one times state B – that sort of thing. This may seem mysterious but it is very essential in the theory. The occurrence of interference depends on this. Also Schrödinger's equation explicitly involves the square root of minus one. We would simply not get enough states to agree with observation if we were to use just real combinations.

So, we have this complex continuum arising in quantum mechanics and we have the real continuum arising in the structure of space-time. It always seemed to me that if we are going to get rid of the continuum in one place we have to get rid of it also in the other. So I tried to develop a theory where one builds up the idea of space and of quantum mechanics simultaneously, starting from combinatorial ideas – purely counting ideas to begin with. I always have had the feeling that counting and other combinatorial concepts are more likely to lie at the root of physics than concepts which depend on the idea of the continuum.

So as well as building up space and time from something more primitive you're trying to get rid of the continuum. As I recall, you chose, as your primitive objects, spinors. A spinor, which has a binary- or two-valuedness, is used in the mathematical description of the electron, but also has a fundamental place in relativity theory.

The point is that spinors can be regarded as basic building-blocks. They are more primitive than vectors or tensors, you see. Vectors and tensors are used in many branches of physics and are quite familiar objects. Now a spinor is, in a certain sense, a square root of a vector. So we can go one step further and build up vectors and tensors out of these spinors. Once you have got the idea of spinors (and you really need spinors in order to describe electrons) they provide the additional advantage of having a certain universality, so that you can build the other objects out of them.

Your first model of space used networks of spinors, didn't it?

The basic idea in that model was to use the concept of spin or angular momentum as the primitive physical concept. You do not initially have the concept of a space in the model.

We can consider two particles, for instance, each particle having a definite value for its spin. If these two particles combine into a single object, that object will have another value for its spin. These values, according to quantum mechanics, must be integral multiples of a basic unit, $\frac{1}{2}\hbar$, the spin of the electron. The rules that are satisfied by spin can be put into a purely combinatorial form. This was really the first step, and it is essentially a matter of rewriting the standard formalism. The second step is then to try to use these combinatorial rules to build up an idea of space. Basically the question is how you define, from these purely combinatorial ideas, what you mean by a direction in space. You have no space to begin with, so you cannot say that an object is spinning in a certain direction. What you can do, however, is take another object and try to define the angle between the spin axes of the two objects. Each object could be some conglomeration of particles which together form some system of a well-defined total spin. Then you can define this angle in terms of certain formalized experiments, the results of which can be treated according to a purely combinatorial calculus. Thus, having a concept of angle between spinning bodies, you can use the concept to build up the idea of physical space.

In this particular model you build up the concept of a three-dimensional Euclidean space, although strictly speaking this is not Euclidean space, but merely the *directions* in Euclidean space. All you can get from this model is the directions.

Is it mathematically possible to build up space with the notions of distance from two-valued objects, for example spinors?

I don't see why not. This particular scheme of mine did not lead to the concept of distance, but it was clear that it was not going to by the way that it was set up. My later ideas are meant to take this into account. But also they are meant to do several other things all at once, so they do not just introduce distance. In particular, I had to make the scheme fit in with the ideas of relativity.

The primitive object in the first theory was a spinor, but now you have developed a new mathematical object, a twistor. Could you explain what a twistor is, briefly?

The idea of a twistor is, as the name is supposed to suggest, a type of generalization of a spinor. It *is*, in fact, a type of spinor really, but not an ordinary space-time spinor. The motivation for developing the twistor theory was partly to improve on this model that I have just been describing in order to build up not only distances but also a proper relativistic space-time.

In the case of a spinor, one can make a physical correspondence – the spin of the electron – to the mathematical object. Is there a physical representation of a twistor?

You can get a picture of a twistor in physical terms, namely, as a particle of zero rest-mass, like a photon, or presumably a neutrino, which moves at the speed of light. If you count up the number of degrees of freedom for such a particle, including its spin, polarization, and location as well as its momentum, you find they are eight in number. These eight degrees of freedom can be conveniently represented mathematically as four complex degrees of freedom. One of the basic ideas in twistor theory is that the complex numbers, which as I mentioned arise naturally in quantum mechanics, appear in the theory right in the beginning. Here one describes space-time ideas, also, using complex numbers. So instead of describing our massless particles by means of an eight-real-dimensional abstract space (as I might have done had I chosen to) I use an equivalent four-complex-dimensional space. This space is not the space we live in. That is to say a point in this space does not represent a point of the space we live in, but it represents the entire history of one of these zero-rest-mass free particles.

To most people, the point is the most primitive geometric concept: but in your theory it is a line.

That's right, because the normal picture of a zero-rest-mass particle, if you think in space-time terms, would be a straight line, for example a light ray. Strictly, for a spinning particle, the straight-line picture is not quite adequate but should be replaced by one of a certain twisting configuration. But let's not worry about that refinement.

This four-dimensional complex space can be related to our familiar space-time, and this complex space is built essentially out of lines rather than points.

That is one way of looking at it. You don't have points as the primitive concepts, you have these twistors, or, as you say, lines of zero-mass particles. If you wanted to find a point, you could do it, but you would have to go to a second step in the theory.

General relativity prescribes the geodesics, or paths the particles will travel on; gravitational forces appear as curvature of these paths. How does this aspect of curvature arise in your complex space?

The twistor theory fits in very nicely with special relativity concepts, but at first you meet with stumbling-blocks if you want to fit twistor theory in with general relativity concepts. I think it does lead to some interesting points of view to do with the nature of space-time curvature but I think it would be rather difficult to explain. In fact, it turns out that twistor theory seems to fit in better with *quantum* ideas of curvature than with classical ones.

One of the basic original motivations of twistor theory always was to try to make the quantum mechanics and the space-time geometry fit together in a much more intimate way; recall the use of complex numbers right at the beginning for the space-time descriptions.

Some people who try to join quantum theory and relativity use ordinary ideas of continuous space-time and apply quantum ideas only to the concept of distance between points, but you are doing something rather different.

It certainly is not just doing that, that's quite right. One of the basic things is to get rid of the concept of point as the most primitive concept. When you try to fit general relativity into this scheme you find that you are almost forced to think of it in quantum mechanical terms rather than as classical general relativity. Perhaps I should not say it quite so strongly, because there are more recent points of view which suggest that perhaps classical general relativity can also be fitted into twistor theory more clearly than I had thought previously. But there is the suggestion that it's very much a quantum general relativity which arises naturally in twistor theory. One of the things that happens once you have general relativity and quantum mechanics coming together in twistor theory is that the points in space cease to have precise meaning – they become smeared. Thus you don't even have points to have distances between. You develop quite different ideas of the basic space concepts.

So that, in a sense, space is built in a quantum mechanical way. Would it be true to say that?

That would be the idea, yes. But the theory is still in a somewhat preliminary stage.

I think it's interesting that we have difficulty speaking about this. This may reflect the fundamental level at which you're working, or the present incompleteness of the theory. Do you feel that it would be easier to speak about it if the theory was complete, or is it of necessity difficult to speak about things like this?

I think it's difficult anyway. One doesn't know where the theory is leading; it may lead to a very simple concept that could be said in a few words. But on the whole it tends to involve ideas which are not easy to express, mathematical ideas which are not very familiar to most people.

Including physicists?

I think so.

Would it be true to say that you have begun a program which attempts to find a foundation for relativity theory within quantum theory?

To some extent that's true. It is really a new way of combining quantum mechanics and relativity. In the first instance, it's a reformulation of existing theory. Much of what I do in this connection is simply rewriting established theory in a completely different language. At first sight, you might think that you're not doing anything new, but different things suggest themselves, and when you get stuck the mathematics guides you. You have to think of how to express certain ideas which you take from conventional theory and you hit on a certain formula which describes these ideas. Then this formula allows generalizations in certain very natural ways, whereas the old formalism wouldn't have suggested that at all. So, very often, although a reformulation is doing nothing more than re-expressing the same physical content in a different language, you find that it leads to something very different. The suggestions as to where to go next can be completely different from the ones in the old language.

You are beginning with a different philosophy, a different view of space-time and matter. Conventional quantum theory involves elementary particles which are beginning to look less elementary because they can always be split up. You have quite a different concept of a particle.

Perhaps I should make the point that in twistor theory, although I've been speaking about a twistor as though it were a zero-rest-mass particle, in fact it is really more like the square root of a zero-rest-mass particle. Twistors are definitely not to be identified with actual physical particles. This leads to a new way of looking at actual particles as entities built up from something more primitive, namely from the twistors themselves.

I was speaking about zero-rest-mass particles; for them a one-twistor description will suffice, but there are other kinds of particles in nature. In fact, the particles we're most familiar with are not of zero rest-mass. They are massive particles, and for massive particles a description in terms of two or more twistors is necessary. There are certain very simple particles called leptons – these include electrons, positrons, and μ-mesons – and my present view is that a two-twistor description is appropriate for them. The internally more complicated hadrons, such as protons, neutrons, and π-mesons, would require a three-twistor description at least. Now in ordinary theory we can talk about particles which are built up of other particles, such as quarks or partons, but these are still always particles of a kind. In twistor theory, however, the components of the particles are not particles but twistors. In fact, it is really rather misleading to refer to particles as composed of twistors; the difference lies in the way twistors are used. Their role is more like the role of *points* in conventional theory. And you don't normally talk about particles being composed of points. The points are what you describe the particles in *terms* of. Similarly, in twistor theory, particles are to be described in terms of a certain number of twistors rather than being thought of as composed of these twistors.

In addition to having a theory which explains elementary particles, do you also explain their interactions in a unified way?

That is the intention. All interactions, according to the twistor point of view, would ultimately find expression in terms of basic interchanges of twistors, although that's perhaps a slightly simple-minded way of looking at it.

The new theory you are developing begins with twistors, mathematical representations of zero-rest-mass particles. A theory which involves only massless objects has a very special symmetry or invariance, doesn't it?

Since in twistor theory one *starts* from zero rest-mass rather than finite rest-mass, one is led to consider conformal invariance. This is a type of invariance which is broader than the invariance in special relativity, where you just consider observers in uniform motion to be equivalent to each other. Suppose we envisage only zero-rest-mass particles or zero-rest-mass fields; take photons as an example – these are particles of light. If you consider light on its own, without any other particles, you are talking about the electromagnetic field and this has a larger invariance group than the invariance of special relativity. Perhaps the easiest way to describe this conformal invariance is to note that what photons *don't* have is a scale. You can imagine the space-time to be altered by a rescaling at each point. As far as the photons are concerned, nothing has been altered. This would not be true of massive particles.

Perhaps a good illustration, in two dimensions, would be to imagine the changes in geometry of the surface of a balloon as you inflate it.

Yes. You can blow up the balloon uniformly until it's twice as big, or non-uniformly, so that some parts of it stretch more than other parts. Suppose there's a little circle drawn on the balloon: If it gets stretched into an elliptical shape, that's not a conformal transformation. If the circle remains a circle, that is a conformal transformation.

In two dimensions, the circle must remain a circle when stretched. Now you deal with stretchings in space-time and the extension of that circle would be the light radiating through space and in time, the so-called light cone.

That's right. Instead of talking about little circles what you talk about are light cones. You can imagine the space being stretched at different points by different amounts, but the stretching has to be isotropic; that is to say, so that the light-cone structure is preserved.

So if your space is filled only with a massless field, i.e. light, then it will be conformally invariant and not possess any scale.

You conformally stretch your space by different amounts at different points and the electromagnetic theory is insensitive to such transformations so there's no way of telling what the local scale is purely by electromagnetic means. There is no way of telling distance.

The twistor theory is based, in the first instance, essentially on this type of geometry, which is insensitive to the stretchings; the concept of distance is not put in right at the beginning. That is, you *can* put the distances in, there's no problem in doing it, but the theory most easily describes these conformally invariant ideas. So, things like electromagnetism, i.e. photons and massless neutrinos, are described very naturally in the theory. Things which involve mass and the breaking of conformal symmetry, though they *can* be put into the theory, don't fit into it at quite the same basic level as do the conformally invariant ideas. Gravitation theory, I should say, is *not* a conformally invariant theory, although it has many conformally invariant aspects, and is, in fact, more conformally invariant than one might have thought.

Has this breaking of conformal invariance something to do with interaction?

Yes, I should think that's fair. The gravitational self-interaction, if you like, where gravitation acts on itself, is not conformally invariant. It *is* sensitive to the scale.

You could have self-interacting theories which are conformally invariant, and those which are not. There's the so-called massless ϕ^4 theory which is conformally invariant, a self-interacting theory which has a very special interaction. On the other hand you can have self-interacting theories, such as Einstein's gravitation theory, which are not conformally invariant, where there's a scale built into the theory.

Now conformal invariance can be removed by certain self-interactions, but the massive particles also break conformal invariance. Do you view mass as produced by the self-interaction of massless objects?

I think one would regard mass more as a result of the interactions of a particle with *all* kinds of particles including itself.

When quantum mechanics first treated the electron, it was interpreted as a particle moving at the speed of light but 'jitterbugging' about so its average motion in any direction was much slower. Is there any connection here with your ideas?

I used to think so, for whenever you observe the velocity of an electron, you find it's the speed of light. The electron can be sitting there not seeming to be going at the speed of light because its jiggling around all the time. You can decompose the wave function for the electron so that it appears as two particles, one flipping into the other and back, and so forth. But, certainly, it would be inaccurate simply to think of this particle jiggling around like that. It's inaccurate for the same reason that all these pictures of point particles are inaccurate.

Could I ask, when you're working on these things, which sound incredibly abstract (the square roots of particles, etc.), do you actually work in a visual sense?

Very much. There are so many things that can be interpreted in geometrical ways, even though they're not geometrical things to begin with. The geometrical mode of thought is a very powerful one. Some people may not agree, and I wouldn't want to claim that one way is necessarily better than another, but it is often very helpful to put a problem, which may not be initially a geometric one, into geometrical language and visualize what's going on. You can often circumvent a great deal of complicated calculations by means of simple pictures.

So although it sounds so abstract to me, you, in fact, have very real pictures.

Yes, apparently concrete pictures. But they are the translations of translations of translations of concepts which may be quite different from the pictures that one ends up with.

What is aesthetics in mathematics and how does it play a role?

I think it's not essentially different from aesthetics in art. It is a feeling for the beauty of the subject and for such qualities as simplicity, universality, and elegance. I think that one's aesthetic judgments in mathematics are very similar to those in the arts. But in mathematics aesthetics is not only an end in itself, but it is also a means. In solving a problem it often turns out that the direct approach – just slugging away, sorting it out – will not get you anywhere, unless you know how to solve it anyway. If you want to find a new way of solving a problem, you must feel your way around, in a sense, and look for pleasing and aesthetically attractive solutions. So in that way aesthetics can be a means towards solving a problem, rather than an end in itself. Of course, it is an end too: one really studies the subject of mathematics mainly for its beauty!

But relativity isn't quite mathematics; it's also part of physics.

Of course. But when you approach relativity from the mathematical point of view these aesthetic criteria are important. Of course, you also want to explain what's going on in nature, and aesthetics comes in there too, because an explanation tends to be successful only if it's pleasing as well.

I wouldn't like to say that the aesthetics are really the only thing in relativity. The subject certainly has a great aesthetic appeal, but any theory must stand or fall by experimental tests. There's no doubt that if experiments did turn against it, the theory would have to be thrown out. It can't survive as a physical theory by aesthetics alone.

John Archibald Wheeler

John Wheeler is a physicist whose mind has the combination of inventiveness and independence that seems so characteristic of American men of talent. Born in Florida in 1911, Wheeler studied with two of the greatest men of science, Albert Einstein and Niels Bohr. Those who knew Bohr say that Wheeler's combination of courteous attention and encouragement to others, no matter if they are established scientists or students at the start of a career, is a reflection of his teacher.

In addition to making contributions in the fields of nuclear physics, atomic structure, and relativity theory John Wheeler has encouraged a generation of young scientists to become independent and creative thinkers. A conversation with Wheeler, or the hearing of one of his lectures, is an exhilarating and entertaining experience. Ideas and examples follow one another until, by a masterly sleight of mind, they condense into a coherent picture and a hint at a theory to come. Wheeler has never been afraid to take his ideas to their theoretical conclusions or to probe the fantastic and imaginative theoretical aspects of our universe. It is perhaps not surprising that on the occasion of his sixtieth birthday he was presented by fellow scientists with a copy of *Alice in Wonderland*.

A few years ago John Wheeler took the courageous step of abandoning geometrodynamics, the approach to relativity theory which had occupied him for many years. In its place he could not yet put an idea, but simply the 'idea for an idea.' After our interview he told us that he looked forward to the years ahead. He would begin again and explore new scientific paths: there would be so much to discover and much to learn.

DP *I'd like to begin by recalling the title of a book you wrote*, Einstein's Vision. *Einstein and Bohr are two great historical figures behind science today.*

Could you tell us, first of all, what you feel Einstein's vision to be, and how it has been reflected in your own work?

Einstein's vision really goes back to the earlier days of William Kingdon Clifford, who proposed that a particle is nothing else but a kind of hill in the geometry of space. When a particle moves from one place to another, it's just as if a hill in the geometry of space, or a wave on the surface of water, moved from one location to another. It was impossible to do anything with an idea so abstract and ethereal until the time came for Einstein's theory of gravity. Now gravity was reduced in its explanation to nothing else but the curvature of the space caused by the sun or any other centre of attraction. With this geometrical picture of gravitation, the door was opened to a much deeper understanding of what goes on, and somehow this genie, which had been introduced solely as a kind of slave to carry force from one mass to another, took on a life of its own. Geometry acquired degrees of freedom of its own; the universe, made up of a kind of curved-up sphere of geometry, turned out not to be able to sit statically, quiet, but was forced to be dynamic. This was so preposterous that Einstein himself, who had brought this genie into being following these earlier free views of Clifford, tried to cork the genie back up in the bottle by introducing some new term that would prevent the universe from being dynamic and expanding. He called it a cosmological term.

Then, twelve years later, Edwin Hubble, the great astronomer, showed that indeed the universe is expanding. This, then, is the greatest, most preposterous prediction that physics has ever made in any time past, confirmed – fantastic evidence of the predicting power of the human mind. To think of this geometry as not only the framework of the universe, but as even supplying the 'hills' that Clifford had talked about, *that* today is still an unrealized dream. But it's an attractive dream, and part of what I called Einstein's vision.

DP *Einstein really did not have much success in reducing matter to geometry. Matter and gravity coexist in a rather uncomfortable way in his theory. Your own work has been directed to removing matter completely into geometry.*

I was trying every possible lead that I could see open, to pursue further this great dream of Clifford and Einstein; to visualize matter as in some way built out of geometry. The new feature that came on the scene in my own thinking, as contrasted with that earlier work, was the quantum principle. One could say, in fact, that the relativity principle of Einstein and the quantum principle of Niels Bohr are the two overarching principles of the physics of the twentieth century, and without taking both into account one can well believe he can account for nothing.

This quantum principle says that geometry, far from being smooth at very small distances, is instead like the surface of the ocean, which may look smooth to an aviator miles above it, but is seen to be covered with waves as he comes down a few hundred feet above the surface. Then, if he is precipitated into a lifeboat floating on the surface, he sees even those waves breaking into foam, in the same way that at small distances, by the quantum theory, space is predicted to have irregularities in its structure, so that if one gets down to sufficiently small distances the irregularities become so gigantic that it's like the foam on the surface of the ocean from the waves breaking. Space is built of a kind of foam-like structure.

Now this really reverses one's view of where particles fit into the scheme of things. Before this time, one thought of particles in space as really important and the space around as relatively unimportant. But now we come to realize that the amount of disturbance that's going on everywhere in space all the time is so great compared to the extra disturbance which a particle makes by being there that it's quite wrong to think of particles as the natural starting-point for the description of nature. You and I look at the sky and we see this and that puffy white cloud floating here and there; the clouds look like the only thing that's important. Yet when we go to study in more detail, we realize that the water vapour in the clouds is a thousand times more tenuous than the air, and that the proper starting-point for the description of the sky is not the clouds but the physics of the air. In the same way, the proper starting-point for the description of particles is all this activity, all the time and everywhere, throughout space.

DP *When Newton proposed his ideas of absolute space, they were criticized by Leibniz in letters to Samuel Clarke. Leibniz conceived of space in a relational sense as defined by material bodies. Has Leibniz's idea been completely abandoned with this new fundamental conception of space and geometry?*

Of course this idea of Leibniz – that in some sense space is not a thing in itself but is a way in which we summarize our knowledge of the relationships between objects – is really, in one sense, played out in mathematical detail in Einstein's already existing theory of relativity. That is to say, in Einstein's theory the geometry is not something which goes its own merry way, but something which is governed in its curvature and its constitution by the location of energy. Where there's energy, space is curved more. In this sense, we have really a relativity: we have a picture of geometry as tied to the location of matter. However, we are certainly not assuming anything like the Newtonian picture – that space is, so to speak, a God-given perfection standing high above the battles of matter and energy – but that space is itself a dynamic participant in the world of physics.

It's quite a marvellous thing how, in all the long history of mankind, the dream has always been held alight by one or another great thinker that somehow we'll manage to reduce all of existence to a mathematical expression in some very deep and marvellous and beautiful sense. Einstein's picture of geometry has been able to carry us towards this vision: that not only space is geometry, but also matter is geometry. I, myself, today, don't believe that it's a sound principle to think of space as the ultimate building material. I think it's not simple enough, and at the same time not complex enough. It's not primitive enough. But right now the more natural thing to do, I think, is to celebrate how absolutely wonderful it is, and how far it's carried us – this idea of interpreting gravitation as simply curvature of empty space. It's preposterous! The only thing that could be more preposterous is how successful it's been!

PB *Do you believe that this approach eliminates man? Are you trying to find, or was Einstein trying to find, a description of the universe, shall we say, from outside, that excludes man? It looks as if one is trying to find a perspective which one properly associates with another sort of being.*

The one reason above all others that Einstein could never bring himself to accept quantum theory, which he himself had done so much to bring into the world, was his feeling that somehow it denied the existence of an objective world; somehow it seemed to make what happens in the world depend upon we who observe it. This seemed to Einstein in contradiction to the objective spirit of science which we've all thought about for so long. In this sense, Einstein *was* really seeking an objective description of nature, if you want to call it that, one in which man's part in bringing about what happens is put aside. He really *did* object to anything that was not objective.

I can remember, myself, trying to persuade him of the correctness of the quantum principle when one of my students, Richard Feynman, had come up with a still newer and simpler way of seeing the content of it. After spending twenty minutes speaking about it to Einstein, and saying how marvellous it was, and asking 'do you not agree, Professor Einstein, that this makes it much more reasonable to accept the quantum principle?' he laughed and said that he had earned the right to make his mistakes, and that he could not believe that God plays dice. In this he was referring to the fact that in the quantum principle we're instructed that the actual act of making an observation changes what it is that one looks at. To me, this is a perfectly marvellous feature of nature. We had this old idea, that there was the universe out there, and here is me, the observer, safely protected from the universe by a six-inch slab of plate glass. Now we learn from the quantum world that even to observe so minuscule an object as an electron we have to shatter that plate glass; we have to reach in there; we have to put some equipment there and we ourselves have to decide whether we're going to put there something that will measure the position of that particle or something that will measure its velocity, and according to which we do, the future of that electron is changed. So the old word *observer* simply has to be crossed off the books, and we must put in the new word *participator*. In this way we've come to realize that the universe is a participatory universe. The question very much on our minds these days is whether this participatory character of the universe extends much farther than that. Is this just the tip of the iceberg that we've seen at this stage in physics? Is it conceivable that, in order to make sense out of the mysteries ahead, we'll find ourselves forced to recognize the participatory character of the universe in a much deeper way than we now see.

I must say that for my own point of view the biggest single issue that is driving us into these problems is not anything that we've talked of so far, but is instead this prediction that the universe is slowing down in its expansion, which we now know from the most compelling observational evidence. Einstein's theory predicts that that expansion will proceed, the universe will reach a maximum dimension, then it will start contracting and undergo complete gravitational collapse. All the laws of physics that we have are based on the idea that we have a space and time in which a law can live and move and have its existence, but when the universe collapses, the whole framework for every existing law of physics collapses. So we have this preposterous situation that a law of physics predicts that the universe will collapse to a point where the laws of physics no longer hold true. Yet all of us feel that physics is that which has truth and substance beyond all superficial changes in the surface of reality. So how are we to reconcile these two horns of the dilemma: physics goes on and physics comes to a halt. I don't know of any crisis in any branch of science in all of human history which is more full of consequences for our understanding of man and his place in the world and the universe than this crisis of gravitational collapse.

PB *The universe is to end, space-time to collapse, and the laws of physics to cease. Time seems to have entered the discussion in a new way. This time seems almost extraneous to the theory.*

This concept of time enters in a most interesting way, and perhaps one really ought to say a word on the side about the way in which our picture of physics becomes continuously more abstract. There was the view of the ancients that the sun goes around in the sky, and that developed into the Copernican picture with the earth going around the sun. Then Kepler told us that it wasn't circle upon circle, but rather an ellipse. That opened the door to Newton, who told us not to look at the *shape* of the orbit at all, whether it was a circle or an ellipse, but to look at the law of force, the inverse square law of gravity. In the intervening years, we've worked up to three or four more levels of abstraction above that. We think of the track of a particle moving through space as governed by a law, but a law which, in some strange sense, makes that particle follow the best of all possible paths. We have something we call the action, and the particle moves on a track of so-called least action. But we ask: how does a particle have sense enough to compare that track with another track? Then we come to the world of the quantum principle, and it tells us that the particle, or whatever it is we're talking about, moving in this orbit is really described by a wave, and the wave really follows all possible tracks; it smells out which is the best. It's in this way that the strange feature comes about, that the motion is as we see it.

In pursuing these thoughts even further we come to a level of abstraction in talking about the dynamics of geometry where we say: 'Look, we talk about a particle moving in space. Nobody can keep us from also asking – what does space have for its arena, in which it moves?' We've come to realize there is a dynamical object we call super-space, a framework. It's not itself dynamic; it's a framework for the dynamics of space. When we look at what goes on from this point of view, we come to see that the idea of time is not a primordial idea; it's a derived idea. It's even an approximate derived idea; it's not an exact idea. The development of geometry with time, the development of space with time, is only a deterministic way of talking. But in the real world, the quantum world, no such completely deterministic principle or way of talking can be right. There are different tracks, if you will, that the dynamics of geometry can take in super-space, and when one comes to talk of what goes on in these terms, then the very idea of time has to be given up, and with it the idea of 'before' and 'after.' They lose their meaning. They're good only at distances of everyday magnitude, but they are not correct ideas down at the very small level of distances that we think of in this foam-like structure of fluctuations in geometry.

As we pursue the task of understanding our physical world at one level and another, coming to these successive levels of abstractness, we realize that many of the ideas that we take for granted should not be taken for granted, any more than the solidity of matter should be taken for granted. Today nobody would argue that matter is solid. Even though it looks so solid from an everyday point of view, we know that it's primarily the space between moving particles and nothing more. And we're not any more doubtful about the reality of the world for that; the world is just as real as ever. When we talk about time and space as being approximate ideas, we recognize that for everyday purposes they're just as good as they ever were. But if we're looking, as we all surely are, for an explanation of how this physical world is put together, we must be prepared to step to points of view that are still more abstract. We don't know yet in what direction, but we are very much reminded of the words of Sir James Jeans some years ago: 'God is a mathematician.' He was only jokingly referring to the idea that there should be some magic scheme of mathematics in terms of which we can understand all the things in the physical world. It's even conceivable that that branch of mathematics is on deck right now today, just as when Einstein came on the scene with his ideas of curved space geometry.

There had been before Einstein the great geometer B. Riemann, who had come up with the idea that the geometry of space is not God-given, but is part of the world of physics. In the period since 1930 there has been a revolutionary development in the world of mathematics and in the world of logic, the very heart and core of the world of mathematics, associated with the names of Kurt Gödel and Paul Cohen. They've shown that things one previously thought to be taken for granted cannot be taken for granted in the world of logic. At the same time, they've emphasized that logic is the only branch of mathematics that has the power to think about itself. This magic feature may be the indication that in logic we must look for the branch of mathematics out of which, in some as yet unconceived way, the physical world is somehow constructed, as Einstein and Clifford could dream of everything being constructed out of geometry. Their marvellous idea, which kept alight the dream of the Greeks, that somehow this is a world of mathematics, has allowed us to delve to greater depth than ever before. Now, with gravitational collapse, we see that that's not enough, that there must be something beyond, something deeper: a point of view in which collapse is no longer such a catastrophe as it seems today, when geometry is everything. In the past, of course, we've never been quite clever enough to see into nature. We've had many a hard knock along the way that's driven this quantum idea into our heads, against the greatest difficulties of us all in understanding it. It's quite conceivable that we shall have to have many additional knocks before we get these new things into our heads, but to me the opposite is also conceivable. One could imagine a gigantic chasm in the landscape, the depth of which is so great that no one is ever going to be able to clamber down the cliff and get across it and get up the other side. Only someone who runs fast enough, is daring enough, and can leap the chasm at one jump will get across. It could be that the mathematics is there just waiting for us to seize it, and to realize how we can make the jump.

Leibniz put the question so beautifully long ago: why is there something rather than nothing? Imagine that we take the carpet up in this room, and lay down on the floor a big sheet of paper and rule it off in one-foot squares. Then I get down and write in one square my best set of equations for the universe, and you get down and write yours, and we get the people we respect the most to write down their equations, till we have all the squares filled. We've worked our way to the door of the room. We wave our magic wand and give the command to those equations to put on wings and fly. Not one of them will fly. Yet there is some magic in this universe of ours, so that with the birds and the flowers and the trees and the sky it flies! What compelling feature about the equations that are behind the universe is there that makes them put on wings and fly? It must be some unique branch of mathematics, which, once it came to us, would strike us as so simple! We would say 'why didn't we think of that before?' If I had to produce a slogan for the search I see ahead of us, it would read like this: that we shall first understand how simple the universe is when we realize how strange it is.

DP *Certainly we have developed an approach, in this century, in which man's thinking about the universe is formalized; physics is axiomatized and mathematics is reduced to logic. But you seem to be saying something more: that the universe itself is constructed according to the same plan. Is this an intuition, a compulsion, or a philosophy on your part?*

When I bring up the world of logic, it is not to say the obvious – that nobody can talk about these things who doesn't pursue logical odds. That, certainly, we must accept and agree upon. Under those circumstances, of course, the propositions which form the part of the logical discussion are propositions *about* something, e.g. 'statement about A is true and therefore B is true.' No, it's not in that sense that I'm talking about logic. I'm talking about logic in the sense of the nuts and bolts, if you will, out of which the world is made, just as Einstein and Clifford were talking about geometry as the magic building material out of which the world was made. That seemed an absolutely preposterous idea in their times. How can you make something out of nothing? How can you make physics out of nothingness? Yet the fact we were able to pursue it so far showed that it wasn't nonsense. And here we are now.

When in 1722 Daniel Bernouilli proposed that such things as heat and thermodynamics and temperature all go back to the chaotic motion of individual molecules, the notion seemed absolutely preposterous. Of course today we know that if we have very large numbers of things, we find in them strange new features that we never see in the individual thing. In the same way, when we have a proposition in a book on logic

that is short and brief, we make out its meaning perfectly straightfor-
wardly. If, however, we think of a proposition which is extremely long,
which takes three feet of type to write out, and if we think not only of one
such proposition but of large numbers of them (and I refer to propositions written in the everyday standard language of logic), we have great
difficulty making any sense out of it; just as it's most difficult to make
much sense out of the collision of thirteen particles in mechanics. It's
beyond the powers of any simple computer to deal with. Yet, somehow,
when we get to this business of very large numbers of particles, things
once again become simple. And we can ask ourselves when we get to very
long propositions and very large numbers of propositions whether we
shall find in the statistics of such objects a feature or features which we
can identify with the things of our everyday physical world. Let me hasten
to say that I'm not talking here about an idea; I'm talking about an idea
for an idea. We're so far today from having the least notion how to proceed in this region! The primary points I would like to suggest are only
these: that the world of mathematics itself should have the tools that we
need to understand the world of physics, and in this world of physics
nothing that we have so far stands up to the crisis of gravitational collapse,
so far as we can see.

Every law can be transcended. We think of a piece of wood as a fossil
from a photochemical reaction in a tree twenty million years ago, and yet
we know that if we subject it to fire, all the molecules will be reconstituted
into new molecules. We think of a steel watchband as composed of iron
nuclei which are fossils from the thermonuclear reaction in a star at ten or
twenty million degrees of temperature some billions of years ago, and yet
we know that if we get to sufficiently intense conditions like those once
again we can reconstitute this iron and make it into new elements. The
crisis of collapse raises in our minds the question: are the elementary
particles themselves fossils from a still more violent event and still higher
temperature conditions and still greater densities at the very earliest time
of the universe itself? And not only the particles, but even the laws of
physics? One used to think, in the world of chemistry, that the laws of
valence were vital, enduring laws. Then one learned that if one goes to
sufficiently high temperatures or sufficiently high pressures, the concepts
of valence fall down.

One by one our laws of physics have this property: that as we pursue
them further they seem like approximations. First we work to a level
where we can bring things under the rule of law, and then we work for
conditions sufficiently extreme that we can show that that law fails, as
with the law of constancy of atomic nuclei which failed when we got to
radioactivity. Now we're talking of the most violent set of conditions that

we have ever been able to conceive – the conditions at the end or the beginning of the universe itself, when, to our view, everything is reconstituted, not only the particles, but, one can imagine, even the laws of physics themselves.

PB *What kind of beauty do you see in the laws of physics?*

The beauty in the laws of physics is the fantastic simplicity that they have. No man who wants to give a woman something that is a token of eternal fidelity does better than give her a diamond. That diamond is characterized by the marvellous constancy of the angles between the crystal planes, the great perfection. Yet if one heats that diamond up to sufficiently high temperatures it becomes vapour and the crystal planes are lost. In that same sense, we think of the laws of physics as beautiful, yet also valid only in a certain range of conditions. As we pursue the search, we come to have a new idea of beauty, and yet it's a strange kind of beauty because we haven't seen it yet. What is the ultimate mathematical machinery behind it all? That's surely the most beautiful of all!

DP *Is this ultimate machinery the quantum principle?*

I would think of the quantum principle as emerging from it, rather than being added to it. One of the truly decisive features that leads one to question the geometro-dynamic picture of the universe, going back to Clifford and Einstein, is the fact that in it one has the geometry as created by God on day one, so to speak, and then on day two God comes along with the quantum principle and adds it on to this geometry to tell it how to behave. In fact, all that we have learned about the quantum principle suggests that *it* is the primary principle. God created it, if one speaks jokingly, on day one, and geometry and everything else came out on day two. But in the first hour of day one, how did even the quantum principle come to be created, and does that go back in some strange sense to logic?

DP *If Niels Bohr and Albert Einstein were here, instead of Paul Buckley and David Peat, do you think that you would have been able to reconcile their views?*

I'm terribly glad you conclude by bringing up Bohr and Einstein, the two greatest figures of our age. T.S. Eliot somewhere says that no poet or artist (and I'm sure he would be willing to include scientist) of any age stands alone. Each added figure changes the perspective one has on all the others. I'm sure our perspectives on Bohr and Einstein will be revised in the next century or two. One of them stands for the quantum principle and the other for the relativity principle. Which will be ultimately the deeper principle time only will tell. I'm prepared to believe that the quantum principle is the deeper one.

DP *You don't envision, then, a synthesis of the two theories?*

In fact, I would call the quantum principle the Merlin principle. You remember Merlin the magician: you chased him and he changed into a fox; you chased the fox and it changed to a rabbit; you chased the rabbit and it became a bird fluttering on your shoulder. As we've chased the quantum principle, and I would call it now the Merlin principle, in each ten years of its history, it's somehow taken on a different colour, each time growing more magnificent in plumage, more penetrating in meaning, and more comprehensive in power.

DP *Of course, you have to remember what happened to Merlin. He got trapped under a rock and he's still there.* (laughter)

Ilya Prigogine

Born in Moscow in 1917, Ilya Prigogine went to Belgium at an early age and has spent the whole of his scientific career there. He is a professor at the Free University of Brussels and also holds an appointment as director of the Center for Statistical Mechanics and Thermodynamics at the University of Texas at Austin. Professor Prigogine's most important contributions to science have been in the fields of statistical thermodynamics and non-equilibrium thermodynamics. Non-equilibrium processes are characterized by energy flows, fluctuations which may be amplified, and emergence of structures. Structures maintained in non-equilibrium situations are called 'dissipative structures' and are characteristic of chemical, biological, meteorological, and astrophysical processes. The theory of dissipative structures has been developed mainly by Prigogine and associates in Brussels and Austin. Ilya Prigogine also has a deep interest in the philosophical aspects of modern science. He was awarded the Nobel Prize for Chemistry in 1977 'for his contribution to non-equilibrium thermodynamics, particularly his theory of dissipative structures.'

DP *First, might we touch upon statistical mechanics? How could you possibly hope to get a description of nature just by considering individual small particles?*

It is a well-known fact that in macroscopic physics we deal with phenomena which look quite different from those described, let's say, in terms of Newton's equations of classical dynamics or Schrödinger's equation in quantum mechanics. As is well known, Newton's equations are time-reversible: when you change t to $-t$, nothing happens. On the contrary, macroscopic physics involves irreversibility; that is the very meaning of the second law of thermodynamics. I think that quite remarkable progress is now being realized in the understanding of the transition between these two types of description. The fathers of statistical mechanics, L. Boltzmann, J.C. Maxwell, and W. Gibbs, perhaps did not have enough confidence in the idea of irreversibility, as it was very natural at the time to admit that it was a basic property of certain dynamic systems. So they tried to introduce an approximation supplementary to mechanics to express irreversibility. In this way, the idea that irreversibility can be obtained only through a kind of *falsification* of dynamics or through a course-graining, as the P. Ehrenfests called it at the beginning of the century, became quite popular.

The new feature is that we have to take irreversibility much more seriously. We are living in a world where transformation of particles is observed all the time. We no longer have a kind of statistical background with permanent entities floating around. We see that irreversible processes exist even at the most basic level which is accessible to us. Therefore it becomes important to develop new mathematical tools, and to see how to make the transition from the simplified models, corresponding to a few degrees of freedom, which we have traditionally studied in classical dynamics or in quantum dynamics, to the new situations involving many interacting degrees of freedom. And the new feature, which I think is very appealing, is that this transition occurs in a very precise fashion through symmetry-breaking, a symmetry-breaking in respect to the parity of the dynamical operators which can occur only in the limit of large systems.

DP *Could you give an example of symmetry-breaking?*

The equations of motion contain a dynamical operator; in fact, we often work with the 'Liouville' operator. When we replace the Liouville operator L by $-L$, and t by $-t$, these equations do not change. We may call this the Lt symmetry. It expresses the fact that the equations which describe evolution towards the future and to the past are the same. However, when you look at the heat conductivity equations, then you no longer have this type of situation. You have a heat conductivity equation which describes the equalization of temperature in the future. Instead of the Liouville operator L you have the heat conductivity, which is a positive quantity. You can still change t into $-t$, but then you obtain a different equation, which you could call an 'anti-Fourier' equation, in which a distribution of temperature would have been uniform in the distant past and becomes non-uniform in the future. Therefore you have a set of two equations, instead of having a single dynamical equation.

What I am alluding to is that, in the limit of large systems, taking into account causality, the equations of motion also take two limiting forms: one for the description of initial-value problems and the other for final-value problems. Each of these two equations is no longer invariant with respect to the transformation $L \rightarrow -L$, $t \rightarrow -t$.

DP *The term* symmetry-breaking *is also used in situations like magnetism, where the individual particles are completely symmetric yet the magnet itself is magnetized in one direction only.*

Yes, but I think the situation is even more complex here. First of all, to derive this result, we need at the beginning to know what the difference between the past and the future is; we need to speak about initial-value problems. We assume that the physicist who formulates dynamics – it's not important whether he applies dynamics to simple or to large systems – can already *at the start* distinguish between initial-value problems and final-value problems. Therefore, we ascribe to the observers of the world, which we are, a sense of the direction of time. Then we show that it is in this direction that the entropy is increasing. How can we then justify the assumption which we made about ourselves? I believe that in many very interesting questions of modern physics, you have to introduce an explanation which involves both physics and some biological aspects as well. This is a new and interesting development. You could say that the idea of reality, as Einstein had it and was defending it, was the idea of a description of the universe without observers, of a reality detached from men. There was a discussion between Einstein and R. Tagore, the Indian poet, in which Tagore pointed out to Einstein that even if such reality would exist, it would probably not be accessible to us; we would not know how to speak in meaningful terms about this kind of reality. Now you see that, both in quantum mechanics and in the problem of irreversibility, it seems that we come to a type of explanation which involves our situation as macroscopic beings which observe the world. We come to the end of the Galilean tradition as defended by Einstein. We must abandon the possibility of an absolute description. That is exactly the type of situation which I have just described to introduce irreversibility, and it is also the situation in the so-called Bohr paradox. Bohr pointed out that we describe microscopic processes, but we speak about microscopic processes in terms of macroscopic physics; in other words, we have to take into account our situation as macroscopic observers when we want to describe the microscopic world. The type of theory to which we came – with my colleague Leon Rosenfeld, who was a long-time associate of Bohr – in the problem of quantum mechanics is very similar to the type of explanation which I have outlined for irreversibility. In the problem of irreversibility, we start with the direction of time as given to the observer, and we show finally that there is, in the limit of large systems, a consistent definition of entropy. Once entropy is defined we can speak about an approach to thermodynamical equilibrium; therefore we can speak about near-equilibrium situations and far-from-equilibrium situations. We know that far-from-equilibrium situations may give rise to structures, to evolution of structures, to successive instabilities which seem to be a first step in the direction of bridging the distance between non-life and life. Therefore we come to the conclusion that there are objects in nature which have a kind of anisotropy of time built in. Recognizing this in the outside world, we may justify the assumption about ourselves which we introduced at the starting-point.

PB *Doesn't that seem like a vicious circle in a way?*

No, I think it is not a vicious circle. It is *a self-consistent model* of knowledge. The problem of measurement, and the problem of the epistemology of quantum mechanics, belong in my opinion to the same category.

In quantum mechanics, we start with observation: we measure probabilities. We don't measure amplitudes of probabilities; we measure the probabilities themselves. Then, in order to correlate what we measure at various times, we introduce the Schrödinger equation, which deals with probability amplitudes. This is quite a different concept from probabilities. However, there is no contradiction, because we can show that in the limit of large systems we deal with situations which can again be expressed entirely in terms of probabilities, and therefore we justify our initial position in this way.

PB *If you're contrasting Einstein's view of a totally detached observer with this one which is self-consistent, you intuitively feel, or at least I do, that there is something about that system that you will never know.*

That may be, I don't know. The main point is that in this description there is no longer a single fundamental level. Our level, the level of macroscopic beings, is not less fundamental or more fundamental than the microscopic level. The idea of physics was always to look for a single fundamental basic level which would be an absolute level. For example, Newton's description was complete, there was no place for anything else, anything below or above, because essentially nature was simply a collection of particles moving according to the laws of dynamics. That was then the whole truth, a complete description of nature. The view which we reach now is that the microscopic level is reached from the macroscopic one but is in turn conditioning the macroscopic level. Therefore, there is no longer an absolute level of description. There are various levels, all interconnected in a much more complex fashion. Of course, this leads to a great change in epistemology because, before, the idea was to explain the macroscopic through the microscopic. That is partly true even now, but there is also the explanation of the microscopic through extrapolation of the macroscopic concept, and this again belongs to this interconnection of levels.

In this connection it is important that recent work leads to an identification of what may be called the limits of dynamics (classical or quantum). It has often been stated that at least in the non-relativistic approximation the properties of matter can be deduced from classical or quantum mechanics (that is, from Hamilton's equation or Schrödinger's equation). If this were true we would have a *single fundamental level* of description. But this is not true, as the properties of matter involve in addition asymptotic properties holding only for long times and described in terms of thermodynamics and kinetic theory. For such global asymptotic situations the consideration of trajectories (or wave functions) is not sufficient. In classical dynamics this is a consequence of what has been termed by Moser the 'weak stability' concept. Thermodynamics and statistical mechanics begin where dynamics ends. This is really at the basis of the 'dialectical' multilevel description of nature.

DP *Would you then find yourself in sympathy with Piaget's ideas on structure?*

You are completely right. It is quite a remarkable coincidence that the ideas of Piaget, the ideas of Bohr, the recent work which we have done with Rosenfeld and others, all point in the same direction: there is nothing like an absolute 'outside,' because we can only apprehend it through our senses. And there is also no absolute 'inside,' because the 'inside' is conditioned through our interaction with the outside world.

There is something common in Bohr's and Piaget's orientation: knowledge (including theoretical physics) is not 'given,' we have to understand its genesis; Piaget speaks about cognitive equilibrium and Bohr about macroscopic language (as imposed by our situation in the world we describe).

DP *Also, in Bohr's interpretation there is a great reliance upon language, or analysis of language, which he said was of classical concepts – that we couldn't probe deeply. Now, you're taking a slightly different view.*

Yes, Bohr's ideas were largely intuitive. Bohr has said deep things, but in a rather obscure fashion. The reason was that the relation between dynamics and statistical mechanics, as well as the definition of the macroscopic level of quantum mechanics, was very vague. Therefore it was difficult to be precise. But I believe that we have now a precise definition of what the macroscopic level is, to which we have to refer the microscopic one. However, this macroscopic level is not necessarily a classical level, because it may still contain the effect of quantum mechanics. Consider, for example, a litre of liquid helium: it is a macroscopic object, but in spite of that it is not a classical object.

DP *When Bohr talked about understanding, could he have been using a different word than you're using?*

No, I think that his direction of thought was probably about the same. There is something so new and so different from classical epistemology in this approach that I expect that for some time there will be a lot of resistance. The classical attitude was to imply that the important things are the elementary particles; they reveal the basic structure of nature, and the rest are derived concepts.

It appears today that physics is something much less monolithic and therefore also much more open to other problems from other disciplines. This is essential if we want to speak in a concrete fashion of fields such as theoretical biology. There, obviously, contact has to be made between biological observations and concepts of theoretical physics, but perhaps a partly new, extended theoretical physics.

DP *So this is a reaction against what was started by the British empiricists, J. Locke and D. Hume?*

I think we are trying to go beyond the distinction of empiricism and idealism, exactly as we try to go beyond the distinction between reductionism and anti-reductionism. If you take the problem of emergence of structure, you see that you have to go beyond a thermodynamic threshold to have structure, to have organization. Therefore, you can really distinguish between levels. There is one level for which chaotic behaviour is characteristic, and one level in which you begin to have space organization, time organization. It's very likely that biology belongs to this level. Therefore you could say there are levels, and this would be a kind of anti-reductionist attitude. But that is not so, because at the same time we proved that they both belong to the thermodynamic description; they are simply different forms which the macroscopic solutions may take according to whether they're near equilibrium or far from equilibrium. Therefore, I think that the opposition between reductionism and anti-reductionism becomes somewhat meaningless.

DP *But what about the concepts in science which are thought to be simple, while other concepts are derived as overlaid from the simple ones? Are you saying that things like particles may not be fundamentally as simple as arguments like topology?*

Yes. For example, the famous Einstein-Rosen-Podolski paradox shows that individual quantum systems, say particles, do not behave as macroscopic objects, do not behave as systems in the macroscopic sense. B. D'Espagnat, a French physicist, has pointed out very beautifully in his recent work that whatever attitude you take, you cannot attribute the same type of reality or the same type of properties to the microsystems that you would like to attribute to macroscopic objects. That shows, of course, that this idea of simplicity of the microscopic world is probably an idea which is gone and will never come back. It is rather curious to see that at a few moments in the history of physics we were very near to the realization of this idea of reaching the fundamental level of description. Newton's dynamics had precisely the ambition to provide us with this level. More recently, when Einstein was working on his unified field theory, if he had succeeded, that would have been this basic description. Also if the world were made of electrons and protons alone, this again would be the basic description. But every time the attempt has had to be given up.

DP *Is it possible to get simplicity through concepts and relations?*

But the simplicity of the concepts is the simplicity in the macroscopic world. The simple concepts are the concepts which are on *our* scale. It is from our scale outwards that we create all other concepts.

PB *There is something quite fascinating here. If we consider evolution, and the asymmetry that seems to exist in the universe, we in fact turn around in a sense and study the process, going backwards, or following it up. Isn't there a kind of fundamental asymmetry here that goes beyond this self-consistency, because we are in fact products of this time flow, or this tendency towards ordering, so we still seem to be trapped in it?*

We are *participating* in it, and we begin now more and more to understand the meaning of this participation. Of basic importance is, of course, the transition between non-life and life, which is obviously a very complex phenomenon. There are all kinds of instabilities which are to be crossed in this transition. Each of these instabilities gives rise to a different organization. We begin to see, through models, like the model of Manfred Eigen, for example, for the competition between biomolecules, in which way this could perhaps have arisen. This is, of course, a very deep change in attitude. From the start of modern physics, there was an emphasis on permanence, on things which are timeless (e.g. conservation laws, elementary particles which would not decay, Newtonian mass points).

Although it may be a bit of an oversimplification, I may suggest that Newtonian physics is a realization of Platonic philosophy. Since the nineteenth century, we emphasize more and more the importance of time. This was already emphasized in Kant, then much more in G.W.F. Hegel, A.N. Whitehead, and H. Bergson. Finally you see it very much emphasized recently in M. Heidegger. More and more one sees that existence, being, is really tied up to becoming, to time. It's quite an interesting coincidence or confluence, I think, that what we are trying now to work out, essentially, is a physics in which the temporal element is much more important. Self-organization, non-equilibrium situations – these are key concepts which find applications everywhere. Of course this change in emphasis implies the study of new tools such as bifurcation theory, fluctuations in non-equilibrium systems, and so on. We have to replace the classical tools based on dynamics (classical or quantum) by new ones.

PB *Presumably one would like to have in this self-consistent description a theory or an idea of how the organization of matter at such very high levels of complexity leads to something called mind, which in turn studies the system from which it has evolved.*

It is a very difficult question. Let us consider some more elementary aspects. Concepts such as chance and necessity are much more complicated than people believed even a few years ago. For example, in the frame of classical physics, and classical thermodynamics, there was essentially no place for life (and even less for mind!) except as a kind of chance product: at some moment, some molecules began to behave in a strange way, and then this fluctuation was propagated, etc. – and you find in authoritative books the idea that life is essentially a struggle *against* the laws of physics. It's like an army of Maxwell demons working against the laws of physics to produce life. A new satisfactory development is that this duality may be overcome. Self-organization comes in when the system is prepared to have it. This implies that the distance from equilibrium is sufficiently large that the description implies non-linearity and bifurcations. Then you can have self-organization which at equilibrium would appear as a miracle.

An example is instability leading to convection. When you heat a liquid from below, at some point it begins to show a pattern of convection. Now this is an extraordinarily complicated event from the point of view of molecules, i.e. you need 10^{23} molecules or so, going along together for macroscopic times. If you are near equilibrium, this would be impossible. It would be a miracle! But, in fact, it does appear with probability one if the system is sufficiently far from equilibrium. This shows you that the ideas of complexity, probability, and so on are basically dependent upon the circumstances in which they can be realized.

PB *But why should the system move away, or be away, from equilibrium anyway?*

There are many aspects to be considered. We live in a universe which permits strong non-equilibrium situations. We cannot imagine what a world in thermal equilibrium would be like. But we *live* in this world and therefore we have these non-equilibrium situations. In the evolution of polymers and biological molecules, we can find transitions which even *increase* the distance from equilibrium. Since each organization requires a threshold distance from equilibrium, we obtain in this way an evolutionary feedback, in which, with increasing distance from equilibrium, we go to higher and higher levels of organization.

DP *Let's explore the notion of order further. What was the classical view of entropy?*

Entropy means evolution. At the beginning of the nineteenth century the idea of evolution was a very central one. It was discussed in biology, sociology, and philosophy. In physics it appeared through thermodynamics and especially through the second law of thermodynamics. You remember the dramatic formulation due to Clausius: the energy of the universe is constant but its entropy tends to a maximum. The meaning of this evolution in physics was explored first by L. Boltzmann. This is a remarkable example of the influence of biology on physics. Boltzmann wanted to become the Darwin of physics. He showed that evolution meant in physics something quite different from what it meant in biology: it meant the forgetting of initial conditions, going towards the most trivial situation. For example, if you put twenty particles into one box, and ten into another of the same size, and permit communication between the two boxes, then you would likely see, after some time, the same number of particles left and right. That is an example of increase of entropy. Therefore, entropy is an evolutionary trend, but related to a rather trivial aspect of evolution, I would say.

DP *To what extent is the notion of increasing entropy a subjective one?*

I believe that is not subjective at all, or, more precisely, that it is not more subjective than the idea of phase transition. This comes back to what I said at the beginning of this discussion: I believe that modern techniques permit us to define a quantity which is expressed in terms of distribution functions and which evolves in a monotonic fashion towards its maximum value. But obviously well-defined conditions have to be satisfied in order to speak of an approach to equilibrium and a function of entropy. The system has to present a minimum dynamic complexity (it is remarkable that according to a recent result this starts with the classical three-body problem!). Also, obviously the increase of entropy does not apply to individual trajectories (or wave functions). As we stated at the beginning, statistical mechanics and thermodynamics start where traditional dynamics (classical or quantum) breaks down.

DP *I was thinking in terms of the traditional theories of entropy.*

In fact, such a 'theory' was lacking, except for special cases (such as a dilute gas), so you had all kinds of interpretations of entropy. For example, some people would say entropy is related to knowledge, and increase of entropy is increase of lack of knowledge. But, if this were so, this would be a trivial statement, and would probably apply to any kind of situation. That is very unlikely. For example, as I said, if you have particles together, they will finally be uniformly distributed. Well, this is not always true. Look at the sky! You see immediately that the gravitational forces create clusters and planetary systems that have nothing like a uniform distribution. Obviously this is related to some features of the long-range forces. Other people were claiming that entropy could only be understood in terms of cosmology. Still if you solve a many-body problem on the computer with a Newtonian program you see an approach to equilibrium. Therefore a cosmological interpretation of irreversibility in this sense has also to be rejected.

DP *There is even the problem of throwing entropy into a black hole which I read about recently.*

That is a most fascinating problem. Certainly irreversibility and relativity have to be tied up much more closely. We would probably need a relativistic bifurcation theory as well as other tools. I feel not competent to discuss this further at present.

DP *Classically entropy was equated or coupled with the word* order. *Now you're giving a more precise definition of entropy. Does this mean that you have a more precise concept of what order means?*

There is indeed a close connection between entropy and order (or disorder!). For an isolated system in thermodynamic equilibrium the entropy is maximum. This corresponds to a maximum disorder (maximum number of 'complexions'). This statement may be extended to other situations such as systems which are in equilibrium with a thermostat. Then it is the free energy ($F = E - TS$) which is minimum.

Such statements express what may be called the Boltzmann order principle valid for equilibrium situations. It is remarkable how many situations may be treated in this way, but still we need to go beyond such concepts to obtain an interpretation of the 'order' around us.

DP *But could you actually pin down what you mean by order, even what order means in everyday life?*

If you consider a town, you have a simple example of 'order' which is not understandable in terms of Boltzmann's order principle. Here the interactions with the countryside play an essential role. If you were to isolate the town it would decay. This is in contrast with a crystal which could be put into a refrigerator where it would continue to survive without any exchange of energy with the outside world.

Such non-equilibrium structures play an essential role in many situations in physics, biology, and sociology. I called them 'dissipative structures' in contrast to equilibrium structures.

Their formation always includes fluctuations and competition between fluctuations. We may therefore speak of 'order through fluctuations' in contrast with the Boltzmann order principle.

PB *You give an example of convection where you had a large number of individual microscopic entities which were together for macroscopic lengths of time, and flowing through macroscopic spaces.*

Yes, near equilibrium such a flow would be impossible; it would involve too few 'complexions' and would appear as a violation of Boltzmann's order principle. But far from equilibrium this situation becomes possible; the deviation from equilibrium (the flow of energy into the system) is transformed into order and a dissipative structure appears.

PB *You could say that there is a causal explanation to that, which may remove the idea of spontaneous ordering – the fact that you're putting energy into the system. If you include the whole system, it might not be ordered.*

Although the flow of energy appears as the 'cause' of the dissipative structure, this does not mean that we have a deterministic theory. Generally there is a bifurcation point and fluctuation theory is necessary in order to 'choose' one of the possible solutions. It is precisely a very fascinating feature that we now have models which go beyond the old dilemma of 'chance or necessity'; they incorporate both elements of deterministic description (especially *between* instabilities) and of stochastic description (near instabilities).

PB *You speak abut thresholds far from equilibrium. Can you explain what you mean by a threshold?*

The critical value as expressed by the threshold is always mechanism-dependent. What is common is that you need some *finite* distance from equilibrium to produce the instability which leads to a dissipative structure. Let us emphasize again that the behaviour far from equilibrium is *mechanism-dependent*. We need autocatalytic or cross-catalytic mechanisms to induce the instability of the thermodynamic branch. This is in contrast with the universality of near-equilibrium behaviour.

This specificity of far-from-equilibrium behaviour is essential to understand the variety of structure which surrounds us as well as the relation between function and structure.

DP *Speaking again of order, I was thinking of hidden order. There are possibly systems which don't appear to us to be ordered, but may have their order folded in. For example, the hologram, the mathematical analogy of which would be a canonical transformation. It may require only a few steps to reveal the order, although subjectively it appears to be totally disordered.*

You are completely right. This is especially true for dissipative structures where order is related to function. It may then happen that it is easier to see the function than the order. Examples in biochemistry could easily be given. The spatial structure of cells and the distribution of enzymes may appear as disordered as long as the chemical pathways involved in the biochemical cycles are not identified.

Dissipative structures, order through fluctuations, become manifest through coherent behaviour in time (i.e. limit cycles) or space. This type of behaviour may appear at a critical distance from equilibrium.

PB *But you can have precise physical definitions of what you mean by coherence.*

Yes. It is a form of cooperativity. Of course cooperativity is known from equilibrium physics. Well-known examples are ferromagnetism or long-range order in crystals. But with dissipative structures we deal with a new supermolecular form of cooperativity or coherence. We may say that the instability structures the space-time in which the chemical processes responsible for the instability proceed. Inversely the processes then become dependent on the behaviour of the system as a whole. We came to concepts such as 'totality' of the system and its evolution through successive instabilities.

DP *Could you correlate order with simplicity and aesthetics, or is that too big a leap?*

We come to very complex problems related to the question of art and aesthetics. There is a book by R. Arnheim on *Art and Entropy*, recently published by the University of California Press. This book has led to a great deal of discussion. The argument in this book goes as follows: increase of entropy is the basic law of nature; life appears as anti-nature;

therefore, there is a basic opposition between artistic activity and nature.

We see now the fallacy of such arguments. It is only near equilibrium that a system tends to the state of maximum entropy and disorganization. Far from equilibrium we may have processes of self-organization as manifested in dissipative structures.

The conception which I have tried to outline here comes nearer to the Greek concept that it is wrong to oppose art and nature. In this perspective artistic activity is a clue towards the understanding of the way nature works.

I may close this discussion with something I have always found interesting from the point of view of the history of ideas. The ideas of atomism and some of the basic ideas of physics were introduced by the Greek atomists, like Democritus, often for *non-scientific reasons*. Of course, they had no experimental proof for the existence of atoms. The reason stated very clearly by Democritus was to deliver men from fear; to make him feel that he has nothing to fear from mysterious forces, from all kinds of unknown things which were surrounding men. That was a very deep and important step, but it has led to another type of anxiety: that this model of the world in which the basic reality would be these moving atoms would be too simple to recognize ourselves in it. This is perhaps the basic reason for the dichotomy between philosophy and the science which developed in the nineteenth century and still continues. Also perhaps, for many, a reason for disinterest in science, as being not of human relevance. Pascal expressed this in a beautiful way: we are afraid to find ourselves the only thinking, the only organized beings in a world of disorganization, in the world of 'stupid' atoms running around. I think that recent developments give us a more balanced view of things, in which we begin to find our place in a world which is not so random, in which structure exists, in which we cannot only perceive the outside world, but we can also feel how *we* originated from the outside world. Thus we are back to the cyclic structure which we discussed at the start of this conversation.

Robert Rosen, Howard Hunt Pattee, and Raymond L. Somorjai: a symposium in theoretical biology

The participants in this discussion are, with the exception of Paul Buckley and David Peat, theoretical biologists.

Robert Rosen was born in Brooklyn in 1934 and obtained a Ph.D. in mathematical biology from the University of Chicago. He is a Killam Professor at Dalhousie University, but at the time of this interview he and Howard Pattee were colleagues at the Center for Theoretical Biology at the State University of New York at Buffalo. His interest in the mathematics of organization at the biological level resulted in his invitation to the Centre for the Study of Democratic Institutions where he has applied some of his theoretical techniques to the study of human organizations.

Howard Hunt Pattee is a Californian born in Pasadena in 1926. He began his scientific career as a physicist before moving into the field of theoretical biology. His current interests involve questions of philosophy and principle in biology as well as in the mathematical description of systems. He has written on the origin of life and the genetic code, including the problem of defining what is meant by a living system. One of his papers, which was read with interest by quantum physicists, bore the enigmatic title 'Can Life Explain Quantum Theory?'

Raymond L. Somorjai (1938–) was educated at McGill, Princeton, and Cambridge universities, where his initial interests were in theoretical chemistry. He moved to the National Research Council of Canada, where his interests in theoretical biology were stimulated. His concern is with the evolution of complexity and hierarchy in biological systems. In particular he is working on the dynamics of protein folding and the interactions of enzymes.

PATTEE It seems to me that we can characterize two points of view towards biology. One of them – I guess this is J.B.S. Haldane's idea – is that physics is a degenerate form of biology, i.e. biology and physics are separated primarily by complexity. Biological systems are enormously complex physical systems ending up perhaps with the brain, which is just a very complex biological system, a collection of cells, interconnections, and interactions. Then there is the other point of view, that complexity alone is not going to be an adequate concept with which to understand the nature of life. I don't mean that it will require biotonic laws; I just mean that in the development of physical systems with many degrees of freedom and many interactions, there will in fact arise new laws, new ideas and principles that make life understandable. I feel that the evolution of living systems has not tended towards greater and greater complexity necessarily; rather, in the sense of function, it has always led to simplicity. The major function of the brain is actually not to sit around and discuss things like we are doing now, but it is to make decisions – it has to decide whether to fight or run or eat, the very simple essential operations that biological systems have – and the purpose of brain function is to reduce the physical interactions, which are enormous in number, to simple behaviour.

ROSEN I think physics grew out of an attempt to answer certain kinds of questions about the world around us. Now, physics has been very successful in answering those questions because the questions have in some sense been simple enough. Complexity is a very fuzzy word. It is used in many different senses. I feel myself that complexity is in the eye of the beholder. Complexity is not intrinsic to a system in itself, but pertains to the ways in which it interacts with what is around it. A system in an environment in which few interactions are possible will seem simple, and a system in an environment in which many interactions are possible will seem complex. So complexity is really in the capacities for interaction that a system has with what happens to be around it. The idea that you have to have complexity in the sense that you were saying is an idea that runs through biology. In evolutionary theory, for instance, we often argue that you have to have many copies of something in order for evolution to proceed. At the genetic level, the idea is that a lot of duplications of the same piece of genetic material were necessary in order for them to be independently variable, so that entirely new paths of genetic development could take place. In another form, we can assert that you can only get from an insect to a vertebrate, let's say, by allowing multiplication of the number of segments of the insect. If you only had one segment, it couldn't vary without having a lethal effect. If you had two, there is freedom for each to take on new interactive capabilities without killing everything.

Complexity, in the sense of redundancy, is something that runs through biology at all levels, even up to the development of multicellularity, which is the redundant multiplying of cells which are then free to vary. In fact, they vary in such a way that each of the cells by itself would not even be viable. But as part of a group of cells in which functions are maintained and shared, entirely new modes of development are possible, which allow organisms to escape from physics, in a certain sense. It is well known, for instance, that a unicellular organism cannot be allowed to get very large, because the surface would go up as the square of the linear dimension, while the volume available to get metabolites in and out would go up as the cube of the linear dimension, implying an absolute upper limit to cell size. Therefore, you could argue, biological organisms could never get very large. But when you multiply the number of copies of cells, suddenly entirely new modes of organization become possible. In that sense we have clues.

I would like to say one other thing. I think that the idea of metaphor gives us a very important clue as to how to approach biological organization. Why is it that we can recognize certain things as organisms at all? It is not because of their physical structure; it is not because of the details of the way they are put together, but because they share certain properties of organization which run common through the biological world, and *this* is the root of our biological intuition. Now, the fact that so many different kinds of physical systems can adopt an organization which will allow us to recognize them as alive indicates that the physics is somehow 'nonspecific.' There is an enormous degree of latitude in physical descriptions which will nevertheless share the same kinds of organizational features. What I have been trying to do is to characterize these, and then try to work back to determine what sort of physical systems allow these to be manifested, rather than to try to start from a specific physical description and try to infer the organization. So this metaphor plays an essential role. I think it is the crucial concept in biology.

SOMORJAI May I just interject something here about complexity. I agree that there are very many ways of using this word, and it is a muddy one, but to my mind complexity almost axiomatically has to include interactions, because if I talk about a million copies of the same thing which are not interacting at all, that doesn't increase complexity.

ROSEN Yes, but it increases the capability for new modes to manifest themselves.

SOMORJAI Only if you introduce coupling.

ROSEN That's right.

PATTEE Physicists had difficulty interpreting the wave function, I think, precisely because it is not, in fact, a predictive model of the world, unless it is coupled to the measurement. So the whole thing has to be looked at as one system. This is, I think, one of the sources of difficulty in interpretations. The physical world that we live in is treated as if it is in parallel with our descriptions of the world.

PB *Is there not some additional step, which is not necessarily contained within quantum mechanics, to apply to biological systems measuring each other?*

PATTEE The trouble with quantum theory is that it is a model produced by the brain of men and not by the cell, so we have difficulty extending these ideas to the most elementary type of interaction which I like to call a measurement. In other words, I think that the language of quantum mechanics probably will be useful or even essential at the cellular level, but we don't know how to apply it very well. We have a difficult enough time with our very special physics experiments which are not really related to processes of living systems very directly.

ROSEN Physicists ask simple questions, and only use a small fraction of the capabilities for interaction of the systems with which they deal. Within physics itself is enormous potential richness, but only a small part of it has been used. As I say over and over again, what is important in biology is not how *we* see the systems which are interacting, but how *they* see each other. This raises entirely new questions and analyses which have no counterpart at the moment in physics, but which are not unphysical. They are part of physics in a real sense and exploit the true richness that physicists have recognized, at least in passing, in formally describing their own systems.

DP *May I ask you about prediction and anticipation? Do you think that the notion of the arrow of time or irreversibility comes in at this level, for to anticipate is almost to see time as moving and flowing?*

ROSEN Yes, I think this is an essential characteristic of organisms: that they must adopt an arrow of this kind in order to observe themselves and to modify their behaviour accordingly.

PB *And yet at the microphysical level, all processes are reversible presumably.*

SOMORJAI The existence of records makes them irreversible.

PATTEE Yes, because in quantum theory, although the wave function is symmetric in time, if you make a measurement you introduce irreversibility by the measurement and this is what I am saying, that the description of a living system must include the measurement process along with the dynamics.

PB *A 'before' and 'after.'*

ROSEN Right. That gives you an arrow to time.

SOMORJAI Irreversibility arises because you made a record of something, some process. The existence of a record introduces irreversibility, because you don't have a record of the future.

DP *I don't think a record is quite the same thing as having a time flowing. You can have records in which everything that has happened is at the same level; for example all past events could simply be recorded without being temporally ordered and separated past from present. But time itself also includes the notion of a 'flowing.'*

SOMORJAI You need access to the record.

PATTEE Implied in my concept of living systems producing measurements is the complementary process of using that measurement to effect future dynamics. That really is what I mean by *description-construction* system. Description and recording alone are quite useless until that record is brought back into real dynamical time.

SOMORJAI You have to have information retrieval.

PATTEE Right. You have to retrieve, read out, and use the information before you complete the process.

ROSEN So you must have access. This is another deep problem of biology.

PATTEE This is the hardest problem because it is easy to see how records are made. There are many physical processes·in which records are left, if you want to interpret them as records, but the crux of the matter is how you *interpret* them. Any history is, in fact, a record of the past, but how you read it out presupposes an entire model of the world, and a biological system must of course have a very simple model. But what it reads and how it reads the gene, for example, is a property of the genetic code; that's what the genetic code does. How this represents a model of the world is a much more difficult question because we don't understand how this model appeared. It seems to be at the moment a mechanism for reading which has no physical basis; its *origin* has no physical basis. It is totally speculative how such a system would arise.

SOMORJAI I think it might be interesting to enumerate in your mind what you consider necessary conditions for the existence of life.

ROSEN I find this a hard question to answer, because my view of life is a functional view expressed in terms of the sort of processes which organisms manifest independent of the physical substratum which is carrying them out; so in some sense a collection of interacting organisms is itself an organism; it's alive in a sense, and has its own description as an organism quite apart from the fact that the components which built it are alive, just as a multicellular organism has a life of its own, apart from the fact that the cells which comprise it are alive. If you take a functional view of this kind, it becomes much more difficult to state necessary conditions in any kind of precise way. What you have to have, at least in so far as we formalize our intuitions about organisms, are modes of coupling with the world which can be regarded as metabolic; we must have inputs from the world, typical material inputs which supply energy and which provide the capacity for renewing the structure of the organism, whatever it might be. So that's a sine qua non; you have to have metabolic apparatus. And you also have to have a kind of genetic apparatus, something which carries information, which tells how the parts which the metabolic part of the system produces shall be assembled both to renew the substance of the organism and also, as a separate function, to reproduce it. I think anything that we would want to call alive would have to have at least those two basic functions: the function of metabolism and what I call the genetic function.

SOMORJAI And wouldn't you say that life requires an open system far away from equilibrium which is self-regulating and adaptive, aside from the metabolic aspect which is, of course, crucial?

ROSEN I think adaptation comes about as a kind of side-effect of what I have already said, and that is already inherent in the physical substratum. If I were to build an organism out of, say, molecules, it would necessarily have adaptive features because of the particular properties of those physical units that I'm using. The dynamics which takes into account the full richness of the system would necessarily have adaptive properties, but these would be different from the adaptive properties which would be manifested if I built the same kind of organism out of different units, out of cells instead of molecules, in which case I would get a different kind of adaptation, but I would always necessarily get a sort of adaptation as a consequence of the fact that my functions were being carried out by some definite structure. That adaptation comes along as a consequence of what I have already said. An open system is also implicit in what I have already said. The very idea of metabolism involves the idea that things are coming in from an outside. It's far from equilibrium.

PATTEE How far?

ROSEN Again this is an idea I have difficulty with. Typically the systems with which you deal in biology, or any system that you would consider complex in the sense you used before, would have many steady states which it can move between. In fact, this is one way of looking at adaptation – the shift of behaviour from what goes on in the vicinity of one steady state to what goes on in the vicinity of another. When you say 'equilibrium,' what you typically mean, I think, is equilibrium in the physical sense, i.e. the system has come to a state of maximal entropy. In that sense, yes, you're far from *that* equilibrium, but I feel that that is really not very informative because the intrinsic dynamics of the system are not such as are going to carry you to that state anyway. They're going to carry you to some other state.

PATTEE Let me say it another way. Let me begin with a biological definition, and then try to translate that into the physical terminology. Living systems consist of a gene which is translated and read out into the phenotype or the organism. The whole idea of evolution depends on the distinction between the genotype and phenotype, and I'm willing to accept this as the basic requirement for life. The question then is: what are the minimum conditions for a genotype-phenotype system?

Now, in order not to use those same words, I translate them into a 'description-construction system,' and then ask what that means in physical terms. First of all (I suppose this is a philosophical presupposition), I look on the world the physicists think about as being primarily involved with space, time, matter, and energy, and the observables that human beings choose as connected with dynamical equations that relate these observables to space and time. In none of these *non-living* systems do we find *internal* descriptions. At least that's not the way we look at physics; we look at it as autonomous and inexorable motions or processes that go on according to the laws of nature. *We* can describe this system and make measurements on it, but the system can only behave according to the laws of nature. In *living* systems, however, we have the process of internal self-description. They also appear autonomous, but when we describe a living system, we must include the genetic input, which is, in effect, a constraint on the laws of motion. Through the genetic code, there are certain molecules – the DNAS – which exert a very peculiar effect on the dynamics of the system, peculiar in the sense of physics. And I believe that the minimum requirement for this peculiar type of interaction is a self-describing, self-interpreting system; this is the basis of self-replication and hence evolution. Now, the particular living systems that we observe today are regarded by most molecular biologists as the *only* systems, or at least the only interesting systems. I think that this is not the most useful way to look at the nature of life. I think one should ask the question: What are the essential properties of such a system? One reason I suspect that the particular system that now exists is not the most fundamental is that all description-construction systems that I know have an essential arbitrariness about them. Jacques Monod calls this the principle of gratuity, which is a good way to say it, but I prefer just to call it arbitrariness, in the mathematical sense. We label things in mathematics arbitrarily, which means nothing more than we have a definite symbol; something is chosen absolutely, but that which *is* chosen is not really relevant. Now, the real mystery is how, from an inexorable physical dynamics, one could ever achieve an arbitrary self-coding, self-describing system. I would repeat that the essential requirement of life is that it is a self-describing, self-constructing system.

DP *How would you find out if a computer or a machine was alive?*

PATTEE I believe it's useful as a strategy to take the widest possible point of view and say that if a computer could describe itself to the extent that it could also construct itself – in other words, if it had an internal machine shop which, following its own instructions, could construct another computer like itself – at least for the sake of argument, I would consider it to be a self-reproducing machine and alive. It could, in fact, under those conditions, evolve by mutation and natural selection, but I doubt if present society would support such activity.

DP *Would you also make the distinction between reflection, self-awareness, and self-description?*

PATTEE Well, for me, that's getting too complicated too soon. I have a difficult enough time just understanding what I mean by self-description–construction; even at the most elementary level it becomes a very difficult physical problem. In automata theory, in formal terms one can use these same words when one is speaking only of abstract symbolic systems which describe and construct themselves, and there's quite a literature about self-describing Turing machines. But that leaves out at least half the problem. In fact von Neumann mentioned this when he first talked about self-replicating machines. He said the formalization of this problem leaves out perhaps what is the most essential part, and that is the physics, the real time, space, matter, energy relationships that are involved in what I mean by construction. In other words, abstract construction doesn't count.

ROSEN You've reduced things back to computation where everything is easy. You're dealing only in symbols again instead of matter, and the real problem is the matter-symbol problem; the philosophical problem is the question of whether symbols are just physical objects. Here I think it's important to assert what one's philosophical assumptions are, otherwise we may be using the same words but meaning quite different things.

PB *Would you elaborate on your present thoughts on self-description, Dr Pattee?*

PATTEE I think the safest point of view is to take the existing biological code and simplify it. We have at present four nucleotides and twenty amino acids as the basic elements of the system. I see no reason why those numbers are magic, though there may be some very strong reasons for numbers in that range – just as there may be strong reasons for having about twenty letters in the alphabet of many languages. But one could simplify it, I think, to two elements for the description and two elements for the construction, and then ask what the minimum conditions are to execute a construction based on a genetic description. This would require that strings of these two amino acids, let's say, can read out or decode the genes (in which the codons might be a single nucleotide) and that the string of amino acids which reads the code is described by the gene. In other words, self-describing means that the actual constructing mechanism must be made out of parts which are described and which can read their own description. And that's too complicated, in my view, to arise under our present picture of primitive soup with random polymerizations of amino acids perhaps and nucleotides. At least it stretches my imagination to see how this happens.

On the other hand, no one has really carefully looked at the various possibilities here. This is one of the things we're doing in my group at the Center for Theoretical Biology* – trying to establish some measure of the probability that self-coding systems arise.

ROSEN Yes, we've talked about this quite extensively. One of the things that you have to watch out for is the richness of the formal system that you're using. If you have, say, four letters and you consider a random polymerization so that any string can appear, then all strings are equally probable and you have an enormous space to search through. So, a priori, the probability of finding anything useful in that mess is effectively zero. But there may very well be, at least in limited environments, constraints

* At the State University of New York at Buffalo; Dr Pattee has since left this center.

which very sharply cut down or alter these a priori probabilities in ways that we can't predict at the moment.

SOMORJAI I think this brings out an important problem and that's the distinction between discreteness and continuity in the description of things. We know that as children we probably first perceive continuity, and number counting comes later on; yet somehow one feels that discreteness is much more fundamental.

PATTEE Some people feel that continuity is more fundamental. This is, I think, related to the description problem. One characteristic of descriptions is their discreteness, at least if you talk about languages. Even the genetic description is discrete. We think of it as a discrete system although if we wanted to look at it in more detail, it would be continuous perhaps. I think that it may be that biology here will tell us more about the nature of the distinction between continuity and discreteness than the other way around. In other words, if you look at biology as Rosen does – functionally – then what is discrete and what is continuous may be determined *for* us by the property of the brain, so to speak, or the property of the measuring device. After all, this is what determines the fineness of any discrete set. Any measurement process cannot have infinite resolution, and automatically reduces a continuous system to a discrete system. On the other hand, the concept of continuity is certainly fundamental in the physical world. We think of our most elementary behaviour as continuous motion. I feel that the first concept of space and time is continuous rather than discrete; discreteness comes as a later abstraction although it is just my own introspection that tells me that my earliest picture of the world was a continuous picture. Perhaps not; perhaps it was discrete and I just didn't know how to express it.

DP *Do you know if Piaget has done any work on this?*

PATTEE The work I know was done on how children learn classifications and logic. The catch is that he was doing experiments, after all, and so in some sense he had to measure what was measurable, and that I think tended to be discretized by his own classification and logical processes.

PB *Conservation was a major experiment as well.*

PATTEE Some conservation or invariance principles may be learned, but whether this is continuous or discrete in its fundamentals is very difficult to say. How does one know if one doesn't know the process of thought in detail? One can't really say what the underlying model is out of which come our linguistic expressions. Once you get into language, you have a very complex mixture of the continuous and discrete. The written symbols are discrete, but, underlying this, the meanings, the semantics, may have an essential continuity; and this may be one of the difficulties we have in connecting syntax or rules dealing with discrete elements with the meaning which may essentially be continuous.

SOMORJAI One does have the feeling, though, that continuity, at least in this mathematical description of the real world, is a convenience; it is definitely much easier to deal with continuous systems, topological systems, from an explicative point of view than a discrete system, a combinatorial system, where the number of possibilities is enormous. Continuity and topological concepts cut these down, but one wonders whether this is the result of our upbringing rather than something fundamental.

ROSEN I really think that the techniques we use in combinatorics are not the ones that would be used in a biologically significant situation. For instance, if you have even a simple optimization problem and you know that the solution is in a set of ten to the hundredth, you say, well, I have to go through that set one element at a time. There's obviously not enough time from the time the universe was created to go through a set of that size. Yet, if the world is discrete, and if biological organisms do exist in that world, and if they do optimize certain things about themselves, what this means is that the *biological* techniques for finding optimization are not of this character. They use or exploit special mechanisms which are not the ones that we think of when we handle combinatorial problems. This is why pattern recognition by computer is so hard. The only thing that *we* can think of doing is giving the computer a vast set of templates.

SOMORJAI But I would say that perhaps nature works by the penalty method, in the sense that even if it is combinatorial and you have a huge number of possibilities, perhaps the self-description sets up severe penalties for searching the whole space and therefore narrows you down to a relatively limited number.

ROSEN You have to exploit special situations – this is what biological systems do – special regularities which don't show up if you take every element of that enormous space on the same terms as every other element. Again I can't reiterate too strongly that what's important is how the biological systems see their world, and not how we see them.

DP *Roger Penrose, who came to physics from mathematics, wanted to rid physics of the continuum and start off with discreteness and combinatorics and try to derive space/time, as it were. David Finkelstein* also feels there's something wrong about the continuum. So from the point of view of some physicists, the continuum is very suspicious.*

* See footnote on page 134.

PATTEE I think that's true. H. Poincaré used to feel the same way – that the whole idea of infinite sets is in the imagination of the mathematician.

ROSEN That may very well be true, but there is a certain area in which there is a correspondence between the results of mathematical analysis which exploits continuity and topology, and things we observe in the world. That correspondence may be fortuitous and it may not. I remember E. Wigner arguing one time that it was amazing that constructs of the mind, particularly of the mathematical mind, should be so applicable to the physical world. Why was it that it worked out so well?

So there is some aspect of truth in the continuum and I feel that we are tying a hand behind our backs if we refuse to recognize this and let it help us all it can. I feel that these things are hard enough without renouncing things which might be useful.

PB *I have a little response to that statement you quote from Wigner. Without prejudicing the whole question of mind and matter and simple identity theory, it's not surprising to me that the brain can do that because the brain still deals with those particular kinds of laws and it's not at all surprising that there should be fundamental resemblances.*

ROSEN But, again, Howard Pattee has often said that mathematics is not a natural activity of the brain.

PB *I don't know. It seems from simple perception experiments that the brain is already doing complex mathematical transforms.*

ROSEN We *interpret* them as complex mathematical transforms.

PB *Perhaps simply two things side by side are already defining a kind of geometry.*

ROSEN I don't think we disagree. What I say is that there is at least homology between the results of assuming that the world is a continuum and arguing on that basis, and the results of such arguments when compared with what's actually happening in the world. I say, if this is so, let us see how far this goes, let us use this homology to the extent that we can. And I say we need all the help we can get.

PB *What is even more beautiful though, and perhaps even more revealing, more important to ultimately explain, is the generation of ideas and relationships and patterns which do not exist at all in any sense in the world, and this is tied in with language very deeply.*

ROSEN That's right. This is something that Danielli, head of our Center for Theoretical Biology, has often said. We start out at what he calls an age of analysis, where we try to explain complex behaviour in terms of inter-action of simpler parts, whatever that might mean, but as soon as we've isolated those parts, we enter a new age which he calls an age of synthesis where we can reassemble those parts in ways which are entirely novel. In biology this manifests itself in the thought that we could actually engineer organisms, put them together out of parts to have preassigned properties, with massive implications.

PATTEE This is the essential difference, in my mind, between events and descriptions of events. One cannot imagine events happening in a way other than the way they do; there's just no way to think of this, whether they're governed by laws or whether they're governed by chance. Bearing my philosophical presuppositions in mind, events have this nature of in-exorability; they are the way they are and that's all we can say. However, descriptions of events can in fact match the events or they can describe or construct things that do not exist. In other words, they can be wrong.

ROSEN We can *create* events.

PATTEE I prefer M. Polanyi's idea that we can 'harness' events. Once one has a description, and if one can read it or construct what is read, one essentially harnesses events in a way that physics does not allow, unless you consider the measurement problem. The measurement problem is a case where the event is essentially altered by the theory, or by the interac-tion, and I think that that is perhaps the degenerate case of construction, that is, construction under a predictive model.

DP *I think it's an interesting hierarchy here, that we started talking about living systems and then about living systems as self-description, and then we started discussing the modes of description and the language we're using; so we're living beings talking about our own modes of description. This suggests a problem of symbols and language.*

PATTEE There is the point of view that the entire structure of physics is con-strained somewhat by language. The question is, by how much? We were talking about discrete and continuous. Which is influencing the other? Is it the physics that induces the concepts or is it the language? Is it the form of description that we are limited to by our brains or by our biological organ-ization? This is the other point that Rosen and Peat were making, about biology influencing physics. It's very difficult to separate the brain from the rest of the body, and in biological or physiological terms or developmental terms, we know that the nerves develop always in conjunction with the muscles, and we also know from experiments in many forms that a person who is deprived of physical contact with objects cannot develop a proper mental picture of an object. That is, the blind man who can suddenly see really doesn't know what to make of the world unless he can touch it. Of course, children's development depends very strongly on the intimate relation of the physical interaction, the visual interaction, and the mental processes that go on. One can't unravel it at the biological level, so why should one expect to unravel it in this very abstract way in physics? In other words, why should one expect that the language does not suddenly influence the form that physics takes? Here, I think, biology will develop a point of view which may radically alter the interpretation of physics. Now, I don't think it will produce the kind of revolution that, say, quantum theory produced, where the actual dynamics changed drastically, but the interpretation of quantum mechanics is very strange and very much up in the air, and I believe that living systems do, in fact, produce this interpretation. Only by understanding how they do it will this interpretation become clear. I'm saying that it is not only a philosophical problem; it is also a physiological problem, or a neurobiological problem. I think many physicists feel that language is crucial for interpretation, that is, the structure of language is more than incidental. Some of these feelings are extensions of Bohr's ideas. However, all these ideas, I think it's fair to say, are not lucid.

ROSEN There's an element of what I could almost call magic in some aspects of discussions like this. People used to argue from the structure of language to the structure of the world: because sentences had subjects and predicates, therefore things had matter and form. To try to argue about the world from the nature of the methods we use to describe the world, this is the essence of magic; that the name of the thing is not accidental but embodies some deep property of the thing, so that if you conjure with the name, you will get some handle on the reality of the thing. Numerology is like this also.

PB *Yes, there were sacred names that you didn't utter.*

ROSEN That's right, very much like science in this way. The name or number that you associate with a physical thing, and a mathematical model that you associate with that thing, are really very similar. In fact the only difference is that we don't think nowadays that the name has any real, deep connection with the thing, whereas we feel that the model does. Some people claim, however, great success with magic, the sort of success that we claim with science.

DP *There may be even great psychological facts in the significance of names.*

ROSEN That might very well be true and this would give magic a scientific basis, if you got the right connection between the name and the thing.

DP *Or the reverse – it would give science a magical basis.*

ROSEN Indeed, many people think it does that.

PATTEE This is one of the reasons that I think that biologists should not treat physics as if it were automatically the fundamental science to which everything should be reduced. That is a kind of modesty in a way, on the part of biologists, and in fact it leads to the idea that once we have translated all biological observations into physical measurements, we have reduced life to physics.

ROSEN This is interesting, because you said earlier that the brain has a fundamental goal – to simplify.

PATTEE Yes.

ROSEN So one often urges that it is imposed upon us by our evolution to take complexity and reduce it to one or a small number of fundamental units or activities.

PATTEE It is, but I think molecular biologists assert that because physical laws are in some way not violated by any experiment, which of course I wouldn't expect, things have been reduced. What I am saying is that, if one moves into physics, one finds physicists sitting around and talking about physics in such a way that one feels that even physics cannot explain itself. So it isn't much of a help to say that this very complicated sort of behaviour is reduced to physics. It's really simpler than physics.

DP *I think quantum theory has shown us that there is no such thing as a unique reduction or a unique description.*

PATTEE Right. That is, it is not an explanation or simplificiation to just say that this set of coding enzymes is obeying the laws of quantum mechanics. That's a very complicated statement; it is not a simple statement; that's what I'm saying.

DP *I think that anecdote about the Rolls Royce would be very useful, as regards reductionism.*

ROSEN There was a Woody Allen routine about him going to England and buying a Rolls Royce. Not willing to pay the duty on it, he dismantled it and packed it in a large number of suitcases. When he came back to the States, he unpacked it and tried to get the Rolls Royce out of that set of parts. The first thing he got was several thousand sewing-machines; then he got a tank, an aeroplane; but he never could recover his Rolls Royce.

This is of the essence in the kind of argument that is involved when I talk about how complex the idea of reductionism really is; how in many ways, unfortunately, as Howard Pattee was saying, biologists want to feel that they have solved all their problems when they isolate a particular set of parts and try to assert that from this set of parts will flow the understanding of everything that they really want to know about life, about organization. What I tried to say earlier was that essentially each way of looking at a complex system requires its own description, its own mode of analysis, its own breaking down of the system into parts; that it is the relation of these different descriptions, which is by no means obvious, which is going to be a source of enrichment not only to biology but also to physics and to all of the sciences up to and including the human sciences. Biological systems are just about the only source of insight we have into organization as such, which will be crucial in handling the kinds of problems which arise at the human level.

SOMORJAI It's not at all surprising that the explanation of certain biological phenomena in terms of accepted and well-known physical principles is all right; after all, we can only make measurements in terms of physical instruments based on the laws of physics; we don't know what kind of questions to ask to separate out phenomena that are not typically of a physical nature. I mean, you cannot carry out experiments.

ROSEN Yes, but the organisms themselves can carry out these experiments.

SOMORJAI But you can't really characterize those.

PATTEE I don't agree. I think we study many structures, many organizations, without using physics. Look at linguistics. We study the structure of symbolic systems without relating it to space, time, matter, or energy. Isn't it possible to study the structure of living systems from this point of view too, without relating it to particular space or time? For example, if you study a biological molecule from the point of view of physics, you get what I would call the structure of a symbol vehicle, just as if you studied marks on the blackboard or a piece of paper in a book from the same point of view. You would get the chemical composition of chalk or graphite or whatever. And this would be, I would say, irrelevant to what was on the blackboard or in the book. The same analogy could work with computers. If you ask what a computer is, you know very well that you get two types of answer, one from the point of view of the computer manufacturer, who says, 'well, a very large number of transistors acting as switches, and many complicated wires and inputs and outputs, etc.' If you talked to a mathematician, he would completely ignore these aspects and talk about computation as a symbolic activity. Only by combining the two descriptions can we approach the reality. I think J. Monod, in his book *Chance and Necessity*, gives an analogy of a man coming from Mars and seeing a computer. Monod assumes that what this man will do, to understand the computer, is take it apart and look at it, part by part. I think this is a bad analogy; in effect, he's using the idea that the computer is basically a physical device. I don't believe he would ever understand the computer very well unless he also understood the nature of computation, which is an abstract idea.

PB *But there is one element that cannot be neglected: we shouldn't look at the question 'what is a computer?' in the same way that we would look at a galaxy or a star or a crystal. We shouldn't, because there is an intermediate step, namely, that we constructed it. And then the Martian may in fact examine the parts, not in order to find out how it itself is constructed, but to determine whether or not it was constructed.*

PATTEE But I say that life is more like a computer than a galaxy. In other words, anything with a symbolic component, anything with an internal description or self-description, cannot be characterized solely in terms of space, time, matter, and energy, in terms of physics. This is why I think the study of life from that point of view is necessarily a half-truth. Just as von Neumann said, the problem of self-reproduction, dealt with entirely symbolically, is a half-truth. This is perhaps one reason why von Neumann's theory never entered biology, because it was a disjoint discussion. Von Neumann, even before it was understood how cells reproduced, before the DNA-RNA protein construction process was known, proposed a logical system which he felt was necessary logically for self-replication. And it was essentially the same as was found in the cell later on. He never got any credit for this from biologists because they said: well, we know the cell isn't made out of Turing machines. That was unfortunate, because I believe an essential part of the theory of biology is the theory of self-describing, self-replicating systems. That's what von Neumann meant by self-replication: evolving cells had to have an internal description. Both of these approaches, I think, are necessary to understand life: the symbolic, logical approach, and the biochemical, physical approach.

PB *They're perhaps not so separate. When you say self-description, and when you say the living system has a model of the world, I presume you would be distinguishing between different kinds of self-description. I mean, a molecule of DNA describing a complex organism is one way, but it needn't describe all possible elements in the world.*

PATTEE No description can completely describe anything.

PB *Okay, so what is your simplest description?*

ROSEN There's a problem here, which arises when you start to talk about descriptions, which complicates this matter. You say that DNA (just to adopt this way of talking) contains a description which is read out to the construction of an organism. If we take a three-body problem, we can't get a description of that three-body system in such a way as to enable us to tell where the three bodies are going to go. The three-body system, however, does have a behaviour, a very definite one.

PB *It seems to know what to do.*

ROSEN Does it have a description of itself? You can get into all kinds of difficult questions when you begin to loosen up a discussion of this kind. Biology is full of traps like this. The basic words of biology all allow many different kinds of interpretations. *Description* is such a word; *fitness* is such a word; *learning* is another.

PATTEE I want to reduce description to a more limited sense. I think that the case of the three-body problem is not self-description; it can be stated that the motion of the three bodies, even though we can't describe it, does not require the assumption of an internal description the way I mean it. It's handled by the laws of nature. It has its own incorporeal laws which exist independently of any description that we happen to give it, which is not the case in living systems. Descriptions must have a physical embodiment; in the cell it is the genetic DNA.

ROSEN What I'm trying to get at, in order to isolate the sense in which we want to usefully use a word like 'description,' is to separate it out from other cases to which the word could be applied but which would only muddy the water. One of the ways of doing this is to construct for ourselves classes of systems in which such words take on definite, precise meanings and to do this perhaps in a number of different ways. In each case, we capture some aspect of the intuitive content of the word that we want to explore.

PB *Can we have some concrete examples of this?*

ROSEN I could give you examples from the standpoint of description, because what Pattee is trying to do precisely is to construct classes of systems which embody a property which he will refer to as description or self-description, and then see what the consequences of that assignment are. I myself, when I was in California in 1972, had occasion to take part in a symposium discussing B.F. Skinner's work, which was really an elaboration of the assertion that every 'behaviour' can be 'conditioned.' Now, 'behaviour' is one of these words. Is this a true assertion or is it not? What I did, although Skinner didn't really think it was constructive, was to create varieties of formal worlds, in which this word *behaviour* took on definite, precise meanings. In each of these worlds you could definitely ask whether every behaviour could be conditioned, if behaviour means this precise thing. Now, these were reasonable worlds; I mean, they weren't completely without intuitive content. In some worlds you could answer the question positively, and in some worlds the answer to the question was no. And indeed, in some worlds, if you conditioned one thing, you could never condition anything else, which would be clearly undesirable. The question became: which of these model worlds is closest to the one in which we live, where we want to be able to assign a definite meaning to this question? That reduces his assertion to something scientific which can be answered. As it stood, the assertion that every behaviour could be conditioned was really quite meaningless. It became a really deep scientific question, then, to decide how to go about approaching the question to get an answer to it. Now, that's a strategy which I think is important to adopt whenever one is considering words like *memory*, *description*, *behaviour*, *fitness*, *learning*, *evolution*, *development* – to try to tease apart the various different senses in which the words can be used and isolate that class of systems which embodies one or another of the intuitive aspects of these words in a formal definite way, so that the consequences can be explored. Now, what Pattee wants to do, as I say, is to isolate a particular kind of meaning of the concept of 'self-description.'

PATTEE First I want to simplify or talk only about sufficiently simple systems so that I don't get mixed up. I think the brain is too complicated. The level of consciousness, so to speak, is too diffuse and complex an area in which to discuss description or self-description in the way I mean it. I feel that it's safe to go back to existing cells because that is what we're trying to explain, and to talk about macromolecules which have specified sequences. This is not unlike higher-level language systems which have alphabets and sequences of letters that form words and which have meanings. The difficulty is in the *meaning*, because one does not normally assume that molecules have meaning. The question is: 'When does a molecule have meaning?' What does that statement mean in terms of physical description? I don't see any logical paradox in this, because it's the same problem that we have in language: 'How does one establish the meaning of a language if one begins with undefined words?' Sometimes the problem of infinite regress arises; you think up a set of symbols and you have to define those symbols and the question is how to define the symbols with which you define the symbols. In fact, this same infinite regress problem was brought up before it was understood how proteins were synthesized. It was thought that enzymes synthesized proteins directly, and this same apparent paradox arose there. You have to have a special enzyme to produce each particular sequence to make one enzyme, and what produces those special enzymes? Obviously this problem is solved by the closure property: it is possible to have a finite set of operations and a finite alphabet which has the self-describing, self-constructing property in the cells; that's one example of this closure, and our languages are another example at a higher level.

ROSEN Let me ask one question which suggests another way in which biology can enrich physics. Where in a word, say in a natural language, does the meaning of that word reside? Certainly not entirely in the sequence of letters, because there are many different words which will carry the same meaning in general, usually expressed by saying that language is highly redundant. There is some sort of kernel subsystem, if you want, of the word, in which the meaning resides, from which the meaning can be extracted.

PATTEE I would look at it the other way. I would look at the word as the kernel itself; that is, the word, the idea of a symbol, is a way of associating a very complex system which a very simple system. The symbol I think of as a simple representation of something very complex. Now, in the cell, let's be specific, in a cell you're asking what is the meaning of the gene for an enzyme, since the gene is the description of the enzyme. First of all, we know that the syntax of the system (i.e., the rules of construction from that description) requires the existence of many other molecules, the transfer RNAs, the messenger RNA, the enzymes, the ribosomes, etc.

ROSEN Yes, it's context-dependent!

PATTEE Exactly; it requires a large set of elements to just construct that enzyme. But the meaning of the enzyme is even broader than that, because once you've constructed it, all you have is a linear sequence. Then it folds up presumably according to an uninstructed set of physical laws and enters the cell as a functional device. The function is its meaning. It may be a lysozyme that chews up the cell, or it may be an enzyme that helps construct the cell. Its meaning is what effects it produces. Meaning, in other words, involves the entire system; one cannot extract the meaning simply; one can only talk about the production of the symbol. In terms of physics, there is a deeper level of meaning perhaps.

ROSEN No. I think what I was trying to say was rather different. This bears on the relationship between the enzyme molecule which carries the active site and the active site itself; the meaning resides in the active site.

PATTEE At that level, yes. But there are many levels of interpretation for all symbolic systems.

ROSEN That is true, but that's another question than the one I was raising. What I was saying was that you can characterize a site *independently*, as a physical system in its own right; it has a physical description which is independent of, or at least related in only a very complicated way to, the description of the molecule which carries the site. Now, many molecules can carry the site. All they need to do is to be able to provide the right observables, to give the site its features, just as to extract the meaning of a word you only need to appreciate certain features of it and not entirely the sequence of letters in it, or anything else about it. Now, this way of talking gives you a new class of systems which are physical but not material. You have physical systems which do not have a separable material embodiment, and that is what I would call the active site of the enzyme. It cannot be extracted from the molecule which carries it.

PATTEE You mean a minimum number of degrees of freedom are necessary to define what the enzyme recognizes.

ROSEN That's right, and those degrees of freedom need not be the ones that enter into the physical description of the enzyme itself. They can be related to them only in a very complicated way, from which we would have great difficulty extracting the function, the site, from the physical description of the structure which carries the site.

PATTEE But isn't ordinary language similar? You're using that as an analogy.

ROSEN That's right. That's what I was just trying to say.

PATTEE Ordinary language theory can be very instructive.

ROSEN But we have to have this capacity for dealing with such subsystems, which cannot be isolated, separated, and characterized, in the ways in which physicists approach material systems. New classes of subsystems are required for this type of analysis.

PATTEE This, I think, is a problem with physics too. I think the measurement problem is of this nature. One cannot characterize a measurement by physical interaction or by complete analysis of the situation, because the measurement really is a context-dependent interaction of the entire device and measuring apparatus, and, in fact, includes the purpose of the observer in order to say whether or not a measurement has occurred. It cannot be stated objectively that it has or it has not occurred without introducing the *meaning* of the interaction.

SOMORJAI Which in a sense means that you require consciousness to measure.

PATTEE Well, I just say 'life' instead of consciousness.

ROSEN I would say 'subjectivity.'

PB *Something bothers me about the word* self-description; *I can't quite put my finger on it. You use the analogy with ordinary language. Now, ordinary language is used to describe not so much the organism that is uttering the language, as to describe its relationship with an outside world. Is it possible to have sets of descriptions in a living organism so that, for example, an enzyme molecule is really describing an environment, some sort of sensing device if you like, and the concatenation of all of these objective measurements leads in some way to what we might call self-description. But in fact self-description is merely a collection of descriptions of the environment as a whole and details in it.*

PATTEE Self-description is a minimum requirement for what I would call a language-like system.

PB *But this is what bothers me, because language is not designed for self-description.*

PATTEE No, but all natural languages have the capability of describing their own grammars, so that one can explain how to use the language in the same language.

PB *But you need a metalanguage, and I think logically the metalanguage is not within the language.*

PATTEE Yes, I think it is in natural languages.

ROSEN This is another example of the difficulties that we can get into if we start using these terms carelessly.

PATTEE Agreed, but the cell can contain paradoxes in a sense too. This idea of a self-destruct system is a paradox; that is you turn on a gene which turns off a gene. Logically, if you write this down, it's paradox, but I wouldn't worry about paradoxes.

PB *Does there exist an analogy to Russell's paradox in biological systems and how does the system resolve that paradox?*

PATTEE Von Neumann made the point that in automata theory, when you introduce real sequences (not real time, but when you have to perform things, let's say, in arbitrary time intervals where there is a before and after), you get rid of many paradoxes, logical paradoxes, because time takes care of that. What I'm saying is that in real physical systems one does not have paradoxes because the topology of space-time takes care of these paradoxes, or at least one doesn't worry about them. The case of contradiction in logic, you see, is a time-independent thing. You prove a paradox by going in a certain symbolic sequence and when you find out that what you said to begin with is wrong, you are upset because it seems to invalidate what you have just said. In a real physical system you simply get an oscillation.

PB *If you feed a paradox to a causal system it converts it into an oscillation.*

PATTEE It's like the box with the switch to open it and a little hand that comes out and turns off the switch and goes back inside.

DP *What about the paradoxes that require you to jump one level in a hierarchy? Do they have a biological analogue?*

PATTEE I'm sure they do. I think that the developmental system is very likely one higher level of description. No one has discovered a way to separate it in terms of the elements, the physical symbol vehicles. But clearly the replication process which involves a single cell is controlled by the developmental process. In other words, cell replication itself is turned off and on by some other part of the gene or the DNA, or at least by some other part of the epigenetic system perhaps. It may not be explicitly stated in the gene 'when you reach this stage, turn yourself off.' There are many procedures which can perform that. There are epigenetic generation procedures in which no explicit string of DNA has to correspond to an explicit action of development.

ROSEN There are several things that are involved here. In physics you take the state description as primitive, and what you are interested in is to talk about the way the system changes state as a result of environmental influences on it. In automata theory one of the things you can do is to define states in terms of equivalence classes of histories, which is the sort of thing you were talking about before. The state becomes an equivalence class of all possible histories going back to minus infinity, as it were, which put the system intuitively into the same state.

PATTEE It eliminates time, the order in time of the system.

ROSEN That's right. So that you have an alternative way of describing what's going on, either in terms of equivalence classes of histories (in which case the inputs are primitive, and the state is a consequence of those), or the state description which physics takes as primitive. Also, the fact that we are dealing with complex systems which admit many descriptions, each of which is only *partially true*, gets rid of many of the logical difficulties that have been raised here, while at the same time raising still others.

PATTEE This is the theory of types in logic; and of representations in biological systems.

ROSEN You see, formal systems have their own properties, their own descriptions as it were, which are not related to the fact that they may model or describe some range of some real activity. The logical paradoxes can arise because of the intrinsic properties of a formal system which don't have anything to do with the fact that it is a model, and since it is only a model over a limited range of activity, the paradoxes which arise in the formalism need have nothing to do with the representation that you use.

DP *Most paradoxes that are like this are pseudo-paradoxes when you move one level higher in the language.*

ROSEN That's also true.

SOMORJAI Is it possible that biological systems develop hierarchical descriptions to move out of paradoxes?

PATTEE That, I think, is very likely. I have a vague idea of how new levels of description can arise, and I suspect that the failure of one level is the basic force or condition for a new level arising. In other words, when a system fails to have a representation or a description to handle a particular situation, it leaves a power vacuum so to speak, or a decision vacuum. I would call it a kind of instability, when a decision needs to be made and there is no decision procedure. One then has ambiguity, and small causes can have large effects. This is, in effect, a crisis in the system, and there can arise then a new type of behaviour.

ROSEN What you were really talking about is just this point of the emergence of novelties in evolution, what the philosophers call a dialectic, the presence of contradictions. Two things contradicting each other, simultaneously coexisting, is an intrinsic property of systems, I feel. It's something that's been a problem in human systems.

PATTEE I would say that it is inherent in the description of systems, not in the systems.

ROSEN No, it is inherent in the systems. We have excluded it from our descriptions and this is why it seems a puzzle. All that you can capture in a description is *part* of the capacity of a system to interact in a system. What you retain has only one of several contradictory aspects which are simultaneously coexisting in that system. This is why you consider the emergent behaviour as unpredictable, simply because you have abstracted away, in your description, that capability which I think is inherent in the systems.

PATTEE We have a hierarchical description problem here. When I say 'description of a system,' you say 'no, the system.' I still say we are talking about a description of the system but at the next lower level. Let's not say 'a system'; I think the very idea of a system is a description. One cannot, in fact, have a system unless one is talking in a language. This is again not an argument; it's sort of a philosophical or intuitive basis for talking about the problem. I believe there is a real world out there for which we do not have a complete description.

SOMORJAI But can you have anything without description?

PATTEE Yes, I *believe* there does exist a world, an external reality. Whether it exists or not is a question we can't answer.

ROSEN It's true, the things that contradict each other are the descriptions.

PATTEE Right.

ROSEN But the interactions which those descriptions describe are there, yes?

PATTEE I agree.

PB *But they are complementary too, aren't they? They are not merely contradictory?*

ROSEN That's right. They are complementary in a sense.

DP *Isn't it probably true that you can't have a perception without some form of description to fit it into? You can't have a perception in a vacuum.*

PATTEE This is almost a matter of definition. What is a perception? You see, physicists use the neutral word *interaction*, and I think most physicists conceive of an interaction as part of the outside world. It's not my mind that produces the interaction; there is something there. The perception of the interaction, on the other hand, is something else.

DP *So when you say there exists an external world without description, I don't quite see what you mean. You certainly can't have any perception of the outside world without coexisting descriptions.*

PATTEE There is no arguing with these presuppositions. If you want to be a solipsist, I can't argue with you.

PB *What about the theory of evolution?*

PATTEE I believe that life arose on the earth about three billion years ago; that's my language. I am being a naive realist. Perhaps it can be solved by looking at it in some form of timeless order as Bohm does in his language, but I don't approach it that way yet.

PB *Another way of putting the problem which is perhaps even more meaningful is (and this again is a Kantian approach): whatever is there (and that includes us, interacting with whatever is there) may not ever be known as though we weren't there, or we may never interact with it in enough ways to get at the full complexity of it. All we can know are those interactions, those differences. I think it is very difficult to go beyond that particular guideline. That is the basic unit of information in cybernetics.*

ROSEN I think that you are now talking in line with the sort of thing that Peat was saying before. It's a slippery area to get into because, in order for systems to be called different, we have to be able to perceive the difference. Can we say, absolutely, that two systems are the same or not? Now, this was one of the earliest things I did in the relationship of physics to, say, biology. Specifically, do we have any basis for resolving the Gibbs paradox? The Gibbs paradox, as you recall, was the following: If you take an enclosure and divide it into two parts, and put gas A in the left side of the enclosure and gas B in the right, and at time zero you remove the partition and let the gases mix, there will be a change in entropy which is independent of the nature of the gases, but dependent only on the fact that they are different. On the other hand, if you put the same gas on both sides of the enclosure, there is no change of entropy. So, Gibbs's idea was that you could let the properties of the gases on one side approach more and more closely those of the gas on the other side and get a discontinuous change in the entropy of mixing when they became identical. Now, how do you get around this paradox? Schrödinger proposed that quantum mechanics could get you around it because it didn't admit that the interchange of identical particles was a real physical event. But I argue that this does not resolve the paradox, because it is still a question to determine whether the particles are identical. This is an observational question, not a theoretical one. How do you know when to apply Schrödinger's principle? Can you settle that with a finite set of observations at your disposal at any particular time? What I did was to construct a variety of formal systems in this case, for which the problem of determining identity was unsolvable in the mathematical sense. It required solving a word problem, and therefore if that could not be solved, then in this context you couldn't even define entropy in an objective way. Now, you see what you were saying here about difference.

PB *I was using it as a primitive definition of information flow, and for perception: that where there is no difference, there is nothing happening.*

ROSEN That we can *see.*

PB *For example, constant velocity is not significant; what counts are accelerations, hence differences.*

ROSEN Those are easy for us to see.

PB *Something travelling along a neural net is really a difference, or a transform of a difference of some kind.*

ROSEN Yes, those are cases where we can actually see what's going on. But there are many situations in which we are not geared to see what's happening.

PB *That's exactly my point – that there are an infinity of differences that are possible and, of these, our own system selects some, which really reinforces the Kantian argument that you never get at that infinity of differences that are possible. Each organism in a very complex environment is, in fact, seeing different things. I tried to imagine one day a bee, a bird, and a man in the same garden, and how this garden was being represented to each of them. You can work it out some way without getting into the full complexities because each has an optical system that operates in certain ranges, etc. But when you try to describe that garden, you'd have to overlay all these descriptions and they are all built on differences. And I think that's all we have – differences that made a difference (to quote G. Bateson). Not all differences make a difference.*

ROSEN They may seem to make no difference now, but there are other environments in which they may make all the difference – this is where the selection comes in, and where variability comes in, and the driving force of evolution comes in. Things which don't make a difference now, equivalent multiplicities which look the same in a particular set of circumstances, serial repetition of sequences, suddenly become important when the environment changes, because then there is a whole new kind of a classification imposed upon them.

PATTEE This is what we ignore in language, I think, and it shouldn't be ignored even at the biological level. In language we make statements that do not have their meaning when we state them, but only assume meaning after we have said them. In fact, I think a lot of speech is this way – we don't know what we're saying until we've heard what we said. We generate *potentially* meaningful statements which only assume their full meaning after they have been stated. This is, in effect, the difference between myth and science. Myth elaborates on the linguistic model in ways which have no corresponding observable correlate, and they therefore have great potential meaning.

PB *They could correspond to psychic components, to psychological components as observables; presumably they are observables.*

PATTEE But when they are stated these are not known. That's what I'm getting at. You see, this is the property of language which is so important in biology: that is, the gene can describe things which do not exist. One does not say this about a physical system.

PB *What about meta-messages? I could be using the same series of words in five or six different contexts and the listener is getting that meta-message which is saying 'this is ironic,' 'this is satirical,' 'this is straight'; how does this apply in your symbolic analogy?*

PATTEE We don't know, of course, whether the gene has any meta-messages of this type, although I would expect it does express itself at many levels.

PB *Have these levels been identified yet?*

PATTEE Not very clearly, unless you want to call the operator and structural genes two different levels, that is, the structural genes say 'make an enzyme with the following sequence' and the operator says 'do not read this message.' That's a higher level. It doesn't care about what the message is, it just says not to read it. Perhaps there is a higher level that says 'until you read this message, do not go to the next gene'; it could be a very complicated hierarchy of interpretations as in any conditional or branching program.

DP *Isn't there such a thing as an evolutionary joke?*

PATTEE Maybe there is, I wouldn't doubt it. Of course, this is why machines cannot translate natural language. We have never been able to build machines which have the property of model-shifting or representation-shifting; this is called the representation problem in artifical intelligence. In order to solve a problem one has to have a good representation. And we build in representation using our own knowledge base. I believe we don't know what we mean by a representation. It's a systems property which has not been defined well enough to program.

SOMORJAI May I raise a slightly different question? I would like to have your reaction to it. What would you say to the statement that our assertion that there are interactions is a consequence of incomplete description?

PATTEE How do you mean? I don't quite understand.

SOMORJAI Well, if you could describe something completely, then the question of interaction doesn't arise.

DP *When you linearize something, then you have a set of separate things that are all interacting, but you can always describe them in terms of maybe just one thing.*

PATTEE Do you mean something like replacing a force by a different frame of reference, by a different geometry, as an intrinsic part of the system? I think that does not remove the essential problem. The idea of force is certainly dispensable, but the language still has to have the property of describing experimental results, and however abstract you make the language, you still have to use the *interpretation* of this formalism to reach an experimental condition. A good example, of course, is classical mechanics, where you begin with the simple idea of force, and then you generalize and abstract and reach the Hamilton-Jacobi theory, when everything is so abstract it takes many layers of interpretation to reach an interpretable or measurable quantity.

ROSEN This is a question which also has an epistemological content. If you take, say, a linear system, where it looks like there are a lot of interactions going on, and reduce it to, or diagonalize, the matrix of interactions, you have then introduced a new set of observables that were not perceived originally. Now you construct them as formal quantities and in that coordinate frame, defined by those new observables which you now take as state variables, they are not interacting; they may as well be bacteria in separate test-tubes.

SOMORJAI This is the point I was raising.

ROSEN That's right. Now, this is again an aspect of the subjectivity that I have been talking about. If you had looked at the system through those new observables originally you would see no interaction, yet the dynamics would be the same.

SOMORJAI This is a point I want to make: that interaction is imposed by our inability to represent something in a proper way.

ROSEN It's not an inability. It's simply empirically convenient for us to look at the original state variables which make the system look interacting.

SOMORJAI Why is that convenient?

ROSEN This is a historical accident, that some interactions are easier for us to engage in than others. We can, and in fact *must*, find the 'right' set of state variables in order to solve our linear system.

SOMORJAI I think it's not a matter of convenience; it's a matter of inability.

DP *I think there is also the philosophical notion of formative cause versus effective cause. If you describe an effective cause then you have several things which interact with each other, whereas describing a formative cause there would be some general form to a thing which would give rise to its behaviour. And you wouldn't necessarily want to split it up into causal relationships between parts of a system.*

ROSEN No, I don't think this is really involved in this question. You can look on the diagonalization of a matrix, at one level, as a mathematical trick to enable you to solve the equations. On the other hand, you can look at it as an experimental matter: we happen to see, or find it convenient to measure, certain quantities about the system which define the dynamics in a particular way. If we had looked at a different set of things, we would have seen the system as if it were not interacting.

SOMORJAI Yes, but your convenience is my inability!

ROSEN There is nothing in principle impossible about having measured those new state variables from the very first. It just so happened; I feel this is a historical accident of what is convenient or easy for us to do – ways in which we find it appropriate to interact with the system.

DP *Would you say a three-body problem in general relativity was a matter of the bodies interacting with each other or an argument by a formative cause?*

ROSEN I can take a three-body system (I think; I haven't done this), and break it apart by defining new observables in just this way, to do for a non-linear system what is easily done for a linear system. At least, I can do it locally, by patching together all these local pieces. I feel it is possible, at least in a wide variety of cases, to decompose any system into non-interacting subsystems, to diagonalize it essentially. The observables in question will look very unnatural to us.

DP *Is this decomposition unique? I mean, it is a bit like reductionism again. You might be able to reduce something, but is that the whole story?*

ROSEN It is a question of analysis. As I said, you have to analyse; you have to get simpler elements somehow. Reductionism as we use it in a strict sense simply gives us one particular recipe for constructing subsystems which I have argued is not adequate.

DP *You have cautioned us about reductionism, and yet we want to simplify. Would you elaborate on your method of discussing observables?*

ROSEN What I said was that reductionism offers us one particular way of decomposing a complex system into simpler subsystems. In biology this way has to do with isolating fractions, simpler physical subsystems, looking at those in isolation, and then trying to give back all properties of the original system from which the fractions came. The assertion of reductionism is that this is universally adequate; that these are the only kinds of system decompositions that you ever need to use. What I said is that if you look at new modes of activity, or other kinds of activity in the system, these modes will not be adequate. I gave you the active site as one example. Any kind of activity brings with it a set of observables or state variables which are used to describe that activity and the dynamics which comes along with it. Now, if that dynamics is complex, in order to solve the equations and understand just that activity in isolation, you need to simplify the system somehow. If it's a linear system, the way to simplify it is of course to reduce it to normal form, to diagonalize the matrix. If it's a non-linear system, you can do an analogous thing. It enables you to solve the equations and then transform back to get the time course of the system activity that you were describing. But it does not describe *all* activities, and for each activity you will get a separate dynamics and a separate way of simplifying. So, what I'd say is not that you cannot analyse, but that the form of analysis is determined by the activity that you're trying to understand.

DP *This appears to be relevant for the measurement problem in quantum mechanics.*

ROSEN Yes. If you take something like an enzyme, which you feel is carrying out some kind of a measurement, it's got a particular mode of functional activity which has a description. That description, as I said, is very far from the description of the molecule which carries the site. In other words, the specific catalytic interactions of an enzyme induce a particular kind of description, a dynamics which you can describe and understand. To solve the equations which describe that dynamics, I introduce a new set of observables which split that dynamics up into simple parts, the simple parts being non-interacting, in a sense. Now they will look very unnatural in a formal sense; nevertheless they will have the property of breaking the original n-dimensional dynamics into one-dimensional dynamics and then you can solve the equations and transform back so that your answer is interpretable in the original state variables that were introduced. For any particular activity which involves measurement or which involves interaction of any kind, it is possible to get a description. This description may or may not look 'physical,' in a conventional sense, in that the observables that are involved may or may not be the sorts of things that a physicist would be able to measure directly about the system. In any case, we have a description that may be too complex to approach directly, so we can't solve it in its present form. What we must do then is to try to extract simpler subsystems from that, and the ones which are naturally suggested are the ones which are in some sense one-dimensional. And I feel in quite general cases we can do this. I can show you how.

DP *These observables which you may get in two different ways may have an uncertainty between them. There may be something analogous to the uncertainty principle in quantum mechanics.*

ROSEN That's true. There may well be, but again, in many kinds of situations at any rate, particularly the ones which involve genetics, you can assert that you are only going to be dealing with commuting sets.

DP *But wouldn't it be true to say that you have a metaphor, a biological, living metaphor, for what happens in the quantum measurement process? Or is that pressing it a little too far?*

ROSEN No, I think the quantum measurement process, first of all, manifests itself directly at the microphysical level in such things as enzyme activity in reading out genetic information. And secondly, the process of measurement itself is exactly the same process which biological organisms use at *all* levels whenever they classify things, whenever they recognize things, whenever they divide up their world around them into classes and act on those classes, whenever they simplify, in Pattee's sense. The brain wants to take complex patterns of environmental stimuli and break them up into classes. One class would say 'run,' one class would say 'approach,' one class would say 'eat,' and so forth. Whenever you do that you have an analogue for the measurement problem, or a metaphor for the measurement problem. That's another thing I tried to say: that the same formalism applies.

PB *Aren't you moving into a new determinism?*

ROSEN In what sense?

PB *Well, when you've projected the measurement problem in quantum mechanics right into microphysical process, you're then removing effectively the observer from it. I mean, you are now making what was formerly a problem of subjective interpretation into an objective one.*

ROSEN It's still subjective for the system which is doing the measurement.

PB *But in our terms it's now objective!*

DP *You're observing something observing something else?*

PB *Yes.*

PATTEE The essential feature of the measurement must involve the 'descriptive failure' of the dynamics that is being measured. In quantum mechanics, the measurement problem is not a dynamical process. In the present interpretation the measurement is an irreversible, non-dynamical process.

ROSEN And so it is in the context that I'm talking about.

PATTEE I didn't understand what you meant by decomposing it, so to speak.

ROSEN I'm decomposing a description but I don't necessarily introduce determinism, because my description is incomplete. I'm only describing *one* activity and breaking up the world with respect to that. Now there are other activities, other interactions, going on. This was the source of the dialectics, so the complementarity will come in right there.

PB *Yes, but what I meant by determinism, by the complementarity, now comes in without a large-scale macroscopic observer so to speak. It's now embedded at a microphysical level. So there seems to be a different sort of problem than our measurement problem.*

ROSEN The measurement problem, for instance in the readout of genetic information, or the activity of an active site, involves explicit microphysical processes in the manner which I described. But the end result, the thing that you observe or that the organism observes, is some kind of gross change in the dynamics of a system which has been affected. Some kind of rate has either been increased or decreased.

DP *Would you care to comment upon S. Comorosan's work which suggests that biological observables exist which have not been detected by conventional physical experiments?*

ROSEN Again, I suggested that the kinds of observables that were involved in the action of such a thing as an active site were not the ones that were conveniently measured in physics; that the biological systems saw each other through different eyes than we would use if we were looking at these systems. So, I suggested that there were other observables that were involved explicitly in these biological interactions, biological measurements. What he tried to do was to find some. He took as an experimental system just this enzyme catalytic system, in which rates can be measured very accurately. There was a very simple enzyme system, a sugar and an enzyme which breaks it down. He took the substrate, the thing on which the enzyme acts, and he perturbed it in such a way as not to modify any of the energetic properties of the sugar. He found that there was a marked change in the rate at which the sugar was broken down in comparison with the untreated molecules.

PB *But he didn't know what perturbation he had in fact introduced.*

ROSEN That's right. He knows it was not of a physical observable.

SOMORJAI There's some cynicism, isn't there, in certain quarters about these experiments?

ROSEN Oh, of course, because, as I say, if the observables in question are different from the ones that the physicist uses, sort of orthogonal to them, a whole new world has been opened up. But the point is that this whole world is comprehended in the language of physics. In physics you always deal with a complete set of observables, out of which the physicist pays attention only to one, such as the Hamiltonian. And when you're dealing with systems like an active site, the concept of a Hamiltonian really has no meaning – you're in a whole different world. That world has been obliquely talked about by the physicist, but not exploited. It has to be exploited in biology; biology makes you look at that world.

PB *So you're saying that one should expand physics to embrace biology rather than reduce biology to physics.*

ROSEN This is what's going to happen. This is in the process of happening. I feel that biology will indeed extend physics, rather than cease to exist as an autonomous science by being swallowed up by physics.

PB *Can we jump several levels higher to human systems and consciousness? Will that also extend biology and hence by implication physics? Here I'm thinking specifically of mind-matter type situations.*

ROSEN This we can't say. As Pattee was saying before, we have trouble enough in biology, but it seems like a natural flow. There is this hierarchy; but one thing about the hierarchy is that the same problems continue to reappear at successively higher levels and we have to use our experience at the previous levels in order to know how to approach them at the next one.

SOMORJAI And of course analogies don't always work.

ROSEN Well, I think they're more than analogies. I think that there is a firm dynamical basis; the analogies are not fortuitous.

PATTEE For example, all descriptions are incomplete at all levels, and if they fail at any point the same type of process may take over. This idea that the failure of description gives rise to a new level would be a general feature of hierarchical organization.

SOMORJAI Once you speak about hierarchical organization, you ultimately imply that it's an open system.

PATTEE Right. In fact, that's the nature of hierarchies which is paradoxical, reminiscent of the infinite regress in linguistic definition.

PB *It's like Gödel's theorem as well.*

PATTEE Perhaps it's the same type of thing, that a hierarchical level always implies that a higher level is necessary to analyse the results of the preceding one, but Gödel's theorem only expresses the incompleteness of one formal level.

DP *Could you elaborate on hierarchies, because people often think of a rigid thing like a government?*

PATTEE My concept of hierarchy is very much more limited; it has to do with the alternation between descriptions and constructions, the idea being that at one level we have dynamical systems in the very general sense, that is, state descriptions and rate-dependent transformations between states which we then describe (or define, if you like) by the proper choice of observables. At another level we have rate-independent global or asymptotic structure, or, if you like, singularities or instabilities. One can label these things differently.

SOMORJAI Catastrophes.

PATTEE Yes, catastrophes. This is my use of René Thom's theory of catastrophes. In fact, René Thom and Ilya Prigogine have both suggested a kind of instability theory of hierarchy. But I want to say that whatever single form of description one has, it is incomplete. One description can only have a limited usefulness. And the way the system is defined will, in fact, limit the range of that description of the system. But since it's only a description and not the system itself, there will arise new behaviour, there will arise failure in the description in some singularity or instability. And this must necessarily give rise to an *alternative* description, since no one description works. The two modes of description, which I call dynamic and syntactic, are complementary in the logical sense.

PB *It's a dialectic.*

PATTEE Right, exactly.

ROSEN Inherent in the nature of systems.

PATTEE This is what I mean by hierarchy: it's an alternation of levels of description upon systems. Another way to say it is an alternation of continuity and discreteness. One can think of the external system as being continuous, obeying dynamics and having instabilities in the sense of continuous systems where infinitely small causes have large observable effects. Now, the description, on the other hand, is always a course-grained view of this, a discrete view, because our languages apparently seem to be limited to discrete strings of symbols. So there are bound to be failures of description whenever the fineness of the description fails to take into account the underlying instability. Even in physics I think one can think of instability as a failure of the description. Consider a fluctuation: nothing has been disobeyed, it's just that we haven't taken into account the initial conditions properly, or the degrees of freedom that are really there, except again at the crucial point where infinitesimal fluctuations produce indescribable results. But this is, I think, a property of the failure of description rather than a failure of laws or anything like that.

PB *This is also a basis for the theory of evolution.*

PATTEE Yes. This is what I would say is the crux of the matter in the origin of life, and the evolution of novelty and creative activity. These three things involve, not an optimization within a description, but an essential failure of the description which gives rise to a new level. This is not Darwinian evolution; it's an alternative to that. I don't dispute the existence of Darwinian evolution, which certainly optimizes within levels. This is a second mode of system behaviour.

ROSEN And this is the crucial one. This is the one which involves what we call function change. A system which we are describing as if it could carry out only a single activity with a single set of state variables and neglect all the rest is also capable of interactions which would make it manifest entirely new functions. When those interactions predominate, you get a change of function; you pass from a structure which was originally described in one fashion and which now requires an entirely new description, to an entirely new mode of activity, an entirely new function, a new organ essentially. And this is the basic vehicle by which evolutionary novelties seem to be generated. There's nothing unphysical about them.

PATTEE You often use the word *emergence*, which is an old word, and sometimes considered discredited by Darwinian evolution, although it's still very active conceptually in arguments about whether Darwinian evolution is adequate. I think that there's no need to use the word *emergence* except to point out that this concept has been around for a long time, but up until the ideas of instability theory, catastrophe theory, and the symbol hierarchy theory, no one could say what it meant.

ROSEN Again, the word *evolution* had been around for a long time and it was not useful until a mechanism was produced to show *how* it could occur.

PATTEE We are saying that there *is* a mechanism for producing emergence.

ROSEN And it's intrinsic in the nature of systems of themselves.

PATTEE Whether or not you can find it explicitly related to direct experiment is, of course, the central problem.

ROSEN Well, *we* have trouble; you see the experiments are being performed *for* us. We have trouble finding a set of experiments which would imply the results of the ones which are taking place in nature. It's an interesting thing; Jonas Salk suggested that the evolutionary capabilities of an organism are *unknowable* until stressed, until a particular set of environmental circumstances appropriate to manifest these evolutionary potentialities exists. That's a very interesting assertion. It says there's no other set of experiments we could do on the system, from which we could infer what the results of *these* experiments are going to be. So they are logically absolutely primitive. There's nothing which implies them, which is a remarkable thing.

PB *So there's nothing in the hydrogen atom as such that would lead one to expect human beings, except now!*

ROSEN Right. And as a matter of fact, this was something that my old professor, N. Rashevsky, used to say. Although you could explain life by physics, you could not predict life from physics. And this also has its roots in the sort of deep things that we've been talking about.

SOMORJAI Dr Pattee, we haven't yet discussed your idea of the difference between law and rule.

PATTEE This again is more or less a philosophical and semantical point of view which can be disputed but neither proved nor disproved. I picture the universe as the primitive concept, the primitive elements of the universe being external to subjects or living systems and being governed by what I call physical laws. Now, I don't mean by physical law *our* description. I mean something outside – the real thing. And I say these laws are *inexorable*, *universal*, and *incorporeal* in the sense that the executor of the law is not itself a structure, like the tape machine which is recording my voice is not incorporeal. It's a real machine which is following certain rules. It converts or transduces various energy in the air here, various energy in the tape, according to a rule. That is what I mean by a rule. It's a transduction which requires a *corporeal* body to execute the rule, and has this element of *non-universality* or *ambiguity*, or gratuity as Monod says. In other words, it could be transcribed in many forms, on acetate disks or tapes or in many other patterns which could be coded by the embodiment of a transducer. But these are rules; I say the living system operates according to rules, whereas physical systems, non-living systems, operate according to laws. It's just another way of saying that descriptive systems are constraints; they require structures to execute the rules. These structures in physics would be described as constraints. Now, in fact, they have to be non-integrable constraints, because if they were constraints that could be hidden in the equations of motion which govern the dynamics, they would not be separable from the laws. That is, it would look as if we had just altered the laws of nature if we added integrable constraints. The genetic code is a particular case of structures which execute non-integrable rules. That's another way of saying that a description requires an interpretation or some structure which serves to do what we call interpreting.

SOMORJAI In practice how do you distinguish between what is a natural law and what is a rule?

PATTEE A rule, I think, can be broken, can fail, or could be changed. A law cannot be broken, cannot fail, and cannot change. Another way to say it is that this incorporeal nature of laws is important; that is, symbols cannot exist in a kind of vacuum. I think of symbols as possible only within a physical context of real matter and real constraints. So, what is wrong, in a way, with formal mathematics is the abstraction away from the executors of the rules. Any discussion of life in completely abstract terms is going to leave out precisely that essential element, and automata theory of living systems is limited to that extent. Real space and time must be included here in order to interpret the meaning of symbols; otherwise they diverge from this real world, and become a world of their own, and have no bearing on it.

SOMORJAI I have some conceptual difficulty seeing the distinction between law and rule in this sense: suppose you devised a rule and you find it's working all the time.

PATTEE What's executing the rule? If I made it up, then either I am doing it, or I've had to build a machine which executes it.

SOMORJAI All right, you build a machine then, which executes it. This is the point. This is corporeal because you'll have a machine that does it. But it never fails. You elevate it to a law.

PATTEE I claim that it will eventually fail for several basic reasons. One is that if there is any element of arbitrariness, which is one of the conditions, then the element of alternative behaviour enters in. That is, it becomes a system in which decision-making is required to execute the rule. What is possible in a physical system, I assume, is the only thing that could happen. Alternatives, in other words, do not arise in physical systems. Another reason is that decision-making is physically irreversible, hence dissipative. Fluctuation error or noise will therefore be unavoidable.

DP *When a law is extended into a new domain, does it cease to be a law? I mean, for example, would Newton's laws of motion cease to be laws and become rules?*

PATTEE No. When you say Newton's laws, you're talking about a description. If you want to execute a description of Newton's laws, you have to actually put the equations in a machine or computer or do something to solve them. If you want to predict the outcome of an experiment, in other words, you are working with a description. My concept of law is not a part of the formalism, but it's a part of the outside world, and we never can reach this.

PB *But is it not possible that among the sets of rules, of which there are possibly very many, there are relations or transformations and these are themselves another law?*

PATTEE Logical laws, you mean; things that are bound to be true?

PB *I am thinking about the deep structure of language.*

PATTEE I really feel that language, the description, in which we talk about all these things, is unavoidable. It's just that I want to talk about it in such a way that I don't become a solipsist; I don't want to believe that everything is language, everything is description. I think there is an underlying physical law which we, through interaction with the outside world, are attempting to describe. And we do it more or less successfully for certain ranges of interactions. But we do it only within the context of rules which we impose. We must impose constraints on ourselves in order to have a language. The syntax, you see, must be there before there's anything we can say.

PB *I agree. What I'm trying to do is find out whether all of these different syntaxes, different constraints, different impositions or choices, relate in some way.*

DP *Do you mean that if you view a law, and then you go to one level higher and look at the law again, it becomes just a rule?*

PB *I'm thinking of art, poetry, and music; the fact that you have a large number of rules and different procedures relating to universal themes.*

PATTEE Now I think I see what you mean. There are many ways of describing the same thing, and each description presupposes a set of rules, or constraints. Reality has innumerable descriptions.

PB *Are these various descriptions isomorphisms or homologies?*

PATTEE Some are and some are not. But the homologies are still part of a description. Homologies have to be described to recognize them as homologies.

PB *So you would suggest they were part of another law.*

PATTEE No; however, this is possible, but I don't see how this would be distinguishable from the laws of nature as I see them. In other words, the hierarchies are really in the descriptions; one has to have a new description to have a new level. I can't see a way to escape from this.

ROSEN Art and music and poetry really describe other worlds which throw a new light on our own in a certain sense. They are metaphors, or at least their impact on us is metaphorical. But those worlds don't exist in space and time.

And so, we are left with many questions.

PB *Are they the right ones?*

PATTEE Well, that's the whole question.

ROSEN I have a hunch that at least some of what's been articulated here is in the right direction, and I feel that finding out whether they are the right questions or not can't help but be constructive; even if the questions turn out to be wrong, we've learned something very important in finding that out.

SOMORJAI Yes, we have learned that we are asking the wrong questions.

ROSEN But more than that: we have learned a great deal about the world, and *why* they are the wrong questions.

F. David Peat and Paul Buckley: reflections after twenty years

PB *The questions we posed on the interpretation of quantum mechanics may still stand. Is there anything new, or does the theory stand complete and consistent?*

DP There still remain these several interpretations and for me the problem seems no clearer; there are still difficulties and ambiguities. One of the people we interviewed was David Bohm, whose last book has posthumously appeared (*The Undivided Universe*). Some may not like Bohm's interpretation, but there are still fundamental questions to be answered. Also, at the time we did the radio programs there was a hope for the final unification of relativity theory with quantum mechanics and it looks as if there are very deep questions remaining unanswered about that. Although things on the surface have changed, there have been a whole series of interesting developments which seemed exciting at the time and which have died down somewhat.

Is the Copenhagen interpretation still the official doctrine of the quantum theoreticians?

There is a remark Basil Hiley (a long-time associate of David Bohm) made to the effect that so many physicists come to praise Neils Bohr but end up thinking like Albert Einstein! There is that surface acceptance of the Copenhagen interpretation as the official interpretation, yet people tend to think in very classical ways. So that there is a difference between what people assert and what they actually practice. The Copenhagen interpretation does seem consistent and complete but there are these difficulties and I don't think that they are any closer to being resolved.

Has the emergence of John Bell's theorem articulated what it means or made the situation clearer?

Some people have said that it is the most important result of the century, that John Bell's theorem tells us something about the nature of the universe; even if quantum mechanics falls, the empirically determined observation has to be accounted for. Dirac said in our interview that maybe quantum theory is a passing theory, but you still have to account for this non-classical correlation.

Essentially Bohr had said that quantum theory was complete, but Einstein always objected to that and tried to find counter-examples. Bohr would spend time with the counter-example and pointed out that there was a flaw in the argument. An idea that Einstein came up with was to take a single quantum object and separate it into two; then they would remain correlated. If you did measurements on one, you could make deductions about the other without ever affecting it because these things were too far apart. We know from relativity theory that signals have to propagate with the speed of light but you could make almost simultaneous measurements within experimental limits, so there could be no possibility of one object affecting the other.

And Einstein thought that in that way one could establish some sort of independent reality of two very separate objects. But independent reality is denied in quantum theory essentially because, as Bohr says, it is a holistic theory. Even if objects are separated by a great distance, they are defined by a single wave function, which is in some sense unanalysable. There should be a correlation between them which is not present in classical theory. This correlation just remained an hypothesis; it was untestable. Bohm recast it when he wrote his famous book (*Quantum Theory*) back in the 1950s and then he wrote down his hidden variable theories. When Bohr read those he said that he had seen the impossible done because they seemed to give some measure of credibility to an objective existence for things. Bohr had denied that quantum objects could have an independent existence. But Bell then tried to reformulate his theories in a very tight way and this reformulation has been experimentally verified in many different ways by Alain Aspect in Paris with the most sensitive methods. It is an accepted observation about the way that the universe is that quantum objects are correlated in ways that are not possible classically, leaving aside what Bohm has done, and perhaps we should come to that later.

This conceptual approach seems difficult to discuss.

What it is really saying I suppose is that classical concepts of space and time don't extend too well into the quantum domain, that the idea of separation and interaction just do not apply. I tried to write a book to explain Bell's theorem (*Einstein's Moon*) just for my own purposes and I found it very difficult to do because all the time you encounter the limitations of concepts and language because you want to talk about things like separation and independence and those are all classical concepts for large-scale objects. Bell's theorem is a remarkable result, and if quantum theory disappears you still have to account for that. No hidden variable theory in terms of minute or microscopic particles moving mechanically can ever account for the quantum theory. The exception to this is the sort of theory Bohm proposed.

Is Bohm's theory non-mechanical?

It is in the sense that he has a quantum potential. In any mechanical theory you have objects and forces and pushes and pulls. The sort of potential which Bohm evokes is a non-mechanical one; it is more like a guide or a field of information. It doesn't push or pull; it is a much subtler potential. And it is also a non-separable potential so it accords with Bell's theorem, which says that quantum mechanics is essentially non-separable, that correlations persist even at macroscopic distances. This is really quite amazing and it's quite a mystery.

As you indicated, Alain Aspect of Paris has performed quite a number of experiments which appear to support or sustain the theorem.

Yes. Each time he has proposed an experiment, he and his co-workers have tried to find a hole in the experiment, and then he has redesigned it so that it will be accounted for. He does the measurements several metres apart and incredibly rapidly so that no signal moving at the speed of light could ever reach from one side of the apparatus to the other. So somehow the quantum objects 'know' what's going on without having to send signals, and, therefore show a connectedness though they don't really 'know' in any sense. To say it in a better way, the concepts of space and separation do not really apply at the quantum level. There may be other ways of dealing with the situation. For example, there is Roger Penrose's twistor algebra. The twistors connect very distant objects. In his geometry, what at our level look like distant objects could be connected directly.

If classical concepts of space and time don't fit, then change them. I've always thought that time in physics was a parameter devoid of richness, and perhaps the quantum theory is showing that there has to be a new concept of time.

Yes, time comes in in a rather arbitrary way in quantum mechanics and you're right about parameter inasmuch as this is what Prigogine has been struggling with, trying to bring in time in a dynamic way. The universe unfolds and time is not simply a parameter in the equations.

We have heard that Prigogine has some interesting new results regarding time.

It seems that he has finally cracked the problem of time, and he would feel that that is a major breakthrough. Take irreversibility. You have irreversibility in the macroscopic world which is supposedly a mere statistical effect, and then you have quantum mechanical measurements which are irreversible. What's the connection between the two?

Prigogine would feel that they have to be unified, that there would be some sort of cohesive theory, that there isn't just an assumption, that this is something very fundamental in the universe. So if Prigogine has cracked the problem of time in physics, it is the problem of the arrow of time he's solved; what time is still remains a mystery.

Perhaps at the quantum level there is no time.

Certainly things only happen when a measurement is registered.

That's what I was thinking. We are introducing time perhaps?

Yes, and time is tied to consciousness.

Consciousness of time, whereas maybe the quantum theory has neither consciousness nor time.

The work on the quantum potential which Bohm was working on before he died is essentially about a field of information. Then he began to talk of it as a field of meaning. In researching his letters I gained the impression that right from the beginning he seems to have thought of nature as a living entity, that there is in a sense a consciousness of life in the universe. So the universe is responding in an intelligent way. But Bohm is still pretty far out.

It is my intuition that quantum theory goes far deeper than we earlier supposed. It is quite sturdy, and Bell's work supports this. Do any recent theories such as chaos theory or superstrings bring anything new to the fundamental problems?

When we did the interviews there were two sorts of people: there were those who wanted to think in very fundamental philosophical ways, for example about the nature of time and causality, and others who were attempting to get an account of elementary matter in the shape of mathematical theories. Superstrings looked like a successful attempt at that. People used to talk about getting the ultimate equation of everything. At the time some people hailed superstring theory as the theory of everything. They really felt that it would resolve and unify quantum theory and relativity. This must be ten years ago. Many people worked on that. In a way it has never quite worked out with the success people had hoped for.

The idea began with Nambu, who noticed that there were patterns of elementary particles, patterns of resonances very like the musical scales you get when you pluck a string. The idea was that rather than thinking of particles we should think of these as being the ultimate entity. Heisenberg cautioned us not to talk of ultimate entities. Rather than points we should think of strings. There are real problems with points. If you get down to points in space the theory broke down. You also got infinities. But what if the fundamental units were little strings which vibrated and rotated and these motions were the elementary 'particles.' That was the initial theory and it sounded very elegant, but when the details were worked out there appeared to be a lot of very deep problems, such as infinities and ghosts and all sorts of things. So people said perhaps these strings are much smaller than we thought, down to 10^{-33} centimetres, the smallest things in the universe. Maybe the vibrations and topological transformations of these strings actually create space, time, and matter and energy. That was the basis of the theory, and it looked as if it could explain everything. And it was very successful, but somehow, as with all of these things in the last sixty years, it has never quite worked out, suggesting that perhaps we ought to go back to these deep questions and resolve them.

And chaos theory?

The other thing that burst upon the scene after we did the interviews was complexity theory or chaos theory or, to put it another way, non-linearity theory. That was another whole area which seemed to burst out of physics into biology, sociology, and economics. It is pretty exciting in a way because it is a more integrative attempt to look at the world. It suggests that chaos and order are interrelated aspects, which I think is interesting because it puts elementary particle physics in perspective; there is a lot of other physics going on. At the time of the interviews, the deepest physics was thought to involve elementary particles and less attention was being paid to ideas of complexity. Then chaos theory came along and showed that natural systems are highly interesting.

It is interesting that before Mandelbrot introduced fractals we had had two thousand years of basing our geometry on the Greeks; nature isn't really like that, it is infinitely complex. Chaos theory showed the interconnectedness of things just as quantum mechanics had shown it. Chaotic systems can be so infinitely sensitive that small perturbations will change them. The 'chaos' period was a sort of healthy, exciting time.

Chaos theory doesn't have the counter-intuitive paradoxes of quantum theory.

No, not really; it is a classical theory. What happened was that some of the equations which were very difficult to solve at the turn of the century – although Henri Poincaré did start it all off by looking at the structure of the solar system and finding that it had instabilities – were handled by the development of high-speed computers, imaging, graphics, and the progress made by some Russian mathematicians in solving equations.

After that came the notion of complexity theory. Then ideas of evolution and attempts to make artificial life; all that has become very fashionable. There is a lot of fashion in science, isn't there? After twenty years I am beginning to see that these are stories that our society tells and that there are fashions for certain sorts of stories. When we began this book we talked to people who were really searching for deep truths. Maybe now people are more interested in fashionable ideas; it is hard to find someone interested in deep truths.

Philosophers, I suppose! Dirac suggested that we shouldn't get too involved in speculation about quantum theory because inevitably it will be superseded. Are people abandoning philosophy too?

Well, there has always been a group of hard-edge empiricists in physics who say that philosophy is for people near retirement. We both know, however, that Einstein, Sommerfeld, Bohr, Pauli, and Heisenberg were all very interested in philosophy. Each one felt that it was important. It doesn't seem the same today. I know that when Bohm died people thought that he was the last of a generation. Physics has become very technical and mathematical. Usually mathematics is a tool used by physicists but Whitten, like Newton before him, is doing original mathematics as well as physics. He worked in superstring theory. The physicist is in the forefront of mathematics in this case.

Before, you made the remark that if the concepts of space and time are inadequate for quantum mechanics then why don't we change them, and this is the sort of thing that Whitten and others tried to do: to go back to topological ideas and create new approaches to space and time. He developed a thing called axiomatic field theory, which is a method of trying to explain the quantum theory in a very deep mathematical way. Things are deep and beautiful, but then I remember talking to one of the inventors of superstring theory – Michael Green – and he felt that when Einstein founded relativity he had deep philosophic insights and sought for a mathematical way of expressing them. With superstrings, he said, people had discovered a mathematical way of describing the universe but there was no deep underlying philosophical theory. He said himself that there wasn't any deep compelling reason for the theory, but Heisenberg did say that if you want to understand quantum theory just look to the mathematics, so there we are.

Maybe this is a crisis developing as we reach the end of the century. When we talk about deep ideas we're both old fogeys; we both think deep ideas mean philosophical ideas. But there is a generation who think that deep ideas are mathematics, and of philosophy as just words to go along with it. If you want to understand the theory, you have to understand the mathematics. The philosophy is not that important; it's a sort of window-dressing.

That's similar to an empiricist view again.

I was speaking to Bohm about that and he said: 'If that's true, I don't want to do physics anymore.'

Yes, there is something seductive about mathematics in that it seems to apply so widely: Wigner's point about the unreasonable effectiveness of mathe-

matics. Yet we are still thinking beings for whom meaning itself is a question. I don't think that philosophy is mere words because philosophy is grounded on something else.

I think that it was interesting that Penrose made a very original attempt with the twistor theory. It went so far and has produced a lot of powerful results, which were used in other fields, maybe unpredictably. So mathematics, as well as being effective as you say, is also unpredictably effective. The theory didn't work out as Penrose had hoped. He went on to look at artificial intelligence and wrote *The Emperor's New Mind* and another book, *Shadows of the Mind*, which is an attempt to answer the arguments brought against the first book. One of the points he made there was about the way a mathematician really works. Certain things are reached in a highly intuitive way that can never be deduced by a sort of piecewise logical argument. When we talk of mathematics we should include the insightful creative leaps. How is it that things can be arrived at which can't be proved in finite steps? I don't know if that is a point about the universe or about human consciousness. It is curious that ideas of matter and consciousness are coming back today. That's again another great fashion. There are a lot of conferences on the topic, and a journal recently started in England focusing on physics and consciousness.

Do you want to get into consciousness?

Yes, and also reality, this idea of reality. What is reality?

Well, any kind of question like 'What is reality?' cannot really be answered simply because in a sense reality includes the questioner. It includes the question so that there is always a loop – there is always a paradoxical loop there. I think, however, that human beings define, and in that sense create, reality and also discover it bit by bit. I would like to know how discovery and reality are related.

I think this idea of a loop, or this self-referential interaction, is something we talked a lot about in the earlier interviews. Bohr said even the disposition to make a measurement affects the whole situation, that we are no longer apart from the universe. And that was a nice thing that came out of chaos theory. Chaos theory is a very beautiful description of complex systems. Immediately you try to verify something by making a measurement, you are disturbing the system, and this is analogous to Heisenberg's uncertainty principle. You are always part of the system, you can't abstract yourself from it. It seems that in both the microscopic and the large-scale world we are very much interacting and a part of it and can't obtain exhaustive information. I think that is the exciting thing to come out of chaos theory. There are certain computer models but you can show that as the system approaches very complex or chaotic behaviour, you can never supply enough information for the computer to completely specify the system. So it looks as if Nature is always escaping.

To me it just means a limitation of human simulation of Nature.

Or the idea of knowing enough to define a system.

Eventually you would need an infinite amount of information. There is a nice theorem about weather. Suppose you want to collect data about the weather. Using fractal theory you can show that laying out weather stations on the surface of the earth is always going to be of insufficient fractal dimension to define the weather itself. So in a very fundamental way we can never have enough information to define the weather. One hopes that this has an effect on people: we can't control, we can't define, we can look at trends. Ultimately science has a limitation to it, at both the quantum and the macroscopic levels.

It is not a surprising result. I mean, it has always been true but there has been some sort of weird desire for minutely specified knowledge, which really comes from feelings of insecurity I think. Both John Wheeler and Ilya Prigogine state that we are participating in Nature. That's an extraordinary thing! But anyone who lives in a tribal world knows what that means. So I have a sense that we haven't learned very much.

It is as if maybe we have a lot of hubris, a lot of arrogance, but are coming to face our limitations, which I think is very important. You are right, in a tribal society there is that sense of participation and renewal and obligations; we've forgotten all that side of it. In biology, people like Brian Goodwin talked about a sacred biology and they tried to look back to that responsibility. I remember when we had an interesting get-together with David Bohm and some Native American elders. They discussed chaos experiments. The elders asked David Bohm what an experiment was. Eventually they said: 'Do you mean you create order in the laboratory?' Bohm said: 'That's it exactly.' Then the elders said: 'What about the disorder you create elsewhere? Have you heard of the ethical implications of what you are doing?' Bohm, who was a deep thinker, was taken aback for a moment. It is another way of looking at the universe. We are participants; everything we do has implications and repercussions.

Essays

Paul Buckley:
evolution and quantum consciousness

These notes present reflections on the limits implied by quantum theory if it is taken in the context of evolution. Within the quantum theory, the uncertainty principle expresses a fundamental limitation on knowing the world, especially on previous ways of knowing the world. What does this mean? Coexisting with quantum theory is the theory of evolution, and if one reflects upon the coexistence of these theories, which are so very different, one may arrive at interesting conclusions. The theories have implications for each other in an informal sense, and these implications involve consciousness. The notes attempt to relate what connections are possible and may provide a certain kind of thinking for those interested in meaning in contemporary science.

1. The seeds of life eventually bloom in ever varying situations. This we know from casual description and from scientific observation and explanation. Since the remote origin of the world a very large number of living forms have existed, many of which are no longer present. The theory of evolution explains the origin of some of these forms, using concepts of variation, inheritance, and natural selection. A different and useful way of talking about the theory of evolution is to say that the theory enunciates some of the reasons for the emergence of various species in time as a result of a dynamic reordering of the whole system. It enunciates a kind of world-adaptive homeorhesis (the term introduced by C.H. Waddington) which implies the appearance of new forms or orders and the disappearance of other forms and orders. This theory initially applies to changes to the macroscopic domains of the living biomass and its environments. To this macroscopic theory are conjoined molecular principles which are grounded in quantum theory, although detailed specification in quantum theoretical terms seems out of reach at the present time.

2. The consciousness which enunciates this theory of evolution finds itself embedded in the evolutionary flow at a very complex level. Of course, it is equally consistent to state that consciousness embeds itself in the flow by inventing a theory which links itself and its factual transformations of the world and that this linking is, at first appearance, a linearly temporal one. A major procedure involves the process of reduction of complex entities and relations to simpler entities and relations. The simplest entities and relations, obtained by reduction of wholes to parts – or better, to simpler wholes – or of presents to pasts, all requiring experimental modification of the contemporary world, become the objects and relations of quantum physics, such as particles, waves, energy transformations, symmetry. Interestingly, this process of reduction seems to be one of the phases of evolution, acting in a complementary fashion, and I shall use the term 'involvement' to include the general process of reduction. Evolution generates complex forms from simpler ones, while the term 'involvement' suggests not really the opposite of evolution but one reversing process of great utility nevertheless. (I say great utility, but there is also a corresponding danger in overemphasising the power of reduction at the expense of synthesis, creation, and moral judgment and it can be completely mistaken if time is not properly accounted for.)

3. Humans have become conscious of evolution, and one reason for this is the existence of a record, a natural one. Vast time periods are involved: it is salutary to remember that life has probably existed for 3.5 billion years in an uncountable variety of forms. In any case, it is true to say that the theory of evolution is one of the contents of consciousness, although this does not imply that evolution is a permanent content. It is possible to experience consciousness with other contents or with no content at all, as in some forms of meditation. One can think of consciousness as emerging within life. One must do so. Within some branch of the living, consciousness emerged. Perhaps the problem of the origin of consciousness involves the problem of the origin of life. If you start on the reductionist trail – or that of analysis – however, you must go all the way down that trail. But in so doing you are not demonstrating a reductionism – it is merely to specify composition and that in temporal terms. That is because there are evolutionary steps behind the present composition. That's the direction of my thinking.

4. Physics, in studying elementary entities and relations, arrives at a principle of uncertainty (as first proposed by Werner Heisenberg) with regard to the knowing of those entities and relations, with regard to the knowing of the objective properties of those entities and relations – i.e., of nature at that level of reduction. Nature's objective reality becomes indissolubly merged with our objective reality, itself part of Nature. This merging seems to introduce a note of subjectivity, but that is not the case; the problem lies deeper, as

was clarified by the Copenhagen interpretation. We find ourself faced with a limit. Initially, that limit is seen from our point of view, as a limit placed upon us by the scale differences between the macroscopic and microscopic levels and by the language differences resulting from the different types of experience. There is a great deal to quantum theory, but here I only wish to emphasise the uncertainty principle. According to this principle we are unable to specify the values of certain conjugate variables, such as position and momentum or energy and time, simultaneously with unlimited precision. The product of the uncertainties of the paired variables cannot be less than Planck's constant, which though very small, is not zero. It is the act of measurement which creates this situation, and the theory accounts for the experimental data extremely well when uncertainty is automatically built in. This knowing of the atomic realm is thus both penetrating yet, in a strange way, limited. In classical physics no such uncertainty principle was necessary, but then a classical object has a well-defined trajectory in space and time, whereas a quantum object does not. Quantum objects thus represent a new layer revealed to physicists. Quantum mechanical knowledge is deep and extensive and has, as we know, many practical applications; it is just the fact of the inapplicability of classical concepts which confronts us in a peculiar way. It appears that a barrier has been found on the scientific path to knowledge of the world, or at least a tight limitation. It should be remembered, nevertheless, that quantum theory provides explicit rules for interpreting the observational situation.

Now the theory of evolution implies that a reality, in the temporal sense of not being affected by our knowing and experimenting, did once exist and that reality evolved toward ourselves (not exclusively but as one branch), who later in the temporal flow do observe the 'past' which is now encapsulated in the 'present.' This suggests, to me at least, that within the mind that contemplates the two theories there ought to exist clear and fundamental relations between quantum theory and evolution theory. Together they do imply consequences not necessarily contained within each one separately, or perhaps contradictions might appear if they are not linked in some way. They implicate each other because they each refer to the world as bridges between concepts and experience of irreversible time.

5. What is the relationship between wholes and parts in a temporal sense? Suppose a system consisting of linked or ordered subsystems. These subsystems we may call parts. Analysis of the parts without regard for the whole would likely give indefinite properties of the parts in terms of the whole. The parts may be understood in terms of properties which characterize each one in itself, yet their properties or relations vis-à-vis the whole will not be characterized unless the whole is characterized. This manifests as a limit.

Suppose the part being considered is not only a part of the system existing at present but existed independently as a whole on its own in the past and formed the more complex whole in a time sequence. Then its full properties and relations must include that potentiality as well as those that relate to existing independently.

We human beings are one result of the evolutionary complexification from elementary entities. We are formed of those entities as they concatenated over immense time periods to create the material aspects of our being. This is true either in real terms or in terms of our theory, our own propensity to theorize about our condition. Thus we are the full systems (among other full systems) and the atoms and molecules point toward us. When studying these atoms and molecules as our past objects, now contemporary with us, we find them open and as yet not completely specifiable through the principle of uncertainty which appears forced upon us. We seem them in their potential state, or in one of their potential states, which is ourselves.

6. Heisenberg has considered that the wave function represents a potentiality or observational potentialities. The evolutionary context being considered here is different in that it enlarges the idea of observational potentiality in time. (I note that Weizsäcker concludes that it is not possible to axiomatize quantum theory without time, in the sense of before and after, though it is quite possible to do so without space.) An interim conclusion, then, is that the uncertainty principle is more than a limit on our knowing the objective state of a system; it is a limit placed on our knowing the past in an evolutionary sense. It may also be an indication that the pure past has no content that is unconnected with the present. In a sense, and this will sound unusual, the answer to our question is ourselves and our theoretical results, including the limit. If this is true, it is a necessary limit to the reduction of wholes to parts and also a limit to the reduction of presents to pasts.

Consider the figure:

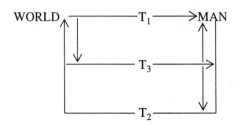

T_1–the World transforms and Man appears.
T_2–Man transforms the World in living/understanding.
T_3–Man, by T_2 into T_1, understands that he is the World transformed.

This looped frame may be found again and again under different guises or in several instances, but here the question being asked is: within this reference frame, where do the limits appear? At first there seem to be two answers. First, looking from the past toward the present, the appearance of quantum theory with uncertainty is an expression of a limit: the world can become aware of itself up to a limit. Second, looking from the present toward the past, the appearance of quantum theory with uncertainty is also the expression of a limit: we can become unaware of ourselves up to a limit. I venture that the limits are the same; there is only one limit.

7. In the figure, T_1 could represent evolutionary process while T_2 represents human efforts to survive and understand the situation, a situation some have called a predicament. Be that as it may, T_3 is a kind of resultant of the efforts wrought by the two transformations. One might also identify T_1 simply with existence, then T_2 becomes reflection. The figure eliminates paradox because it represents ongoing process and not fixed terms. The terms 'existence' and 'reflection' are already dynamic and full of meaning. If one takes existence to include evolution in a deep way, existence then refers to a series of time-ordered states which initially had a given direction and one direction might be complexity, though that would not exhaust the repertoire.

The higher consciousness represented by T_3 is associated with the manipulation of natural process and, today, of life forms in the framework of molecular biology. Not a few troubling ethical issues have emerged or will emerge as a consequence of this manipulative power and devotion to technique, but these will not be discussed here. Suffice it to say that they are closely associated with the particular consciousness which science represents or, more accurately, which a version of science represents. A central aspect of this consciousness is reduction of the complex to the elementary and, as I said earlier, this has an involving character about it. Reduction is a key element in T_2 while the other key element, synthesis, has the character of T_3. T_3 has the character of judgment.

In science, quantum theory emerges at the intersection of two movements: evolution and involvement. Or existence and reflection. The intersection manifests itself as a doubly edged limit. This does not assume any realism. I do not say, for example, that there is an evolutionary process in an objective sense and that we are studying the process (although that is not excluded either); I am considering the operations of our thinking in its theoretical activities as it both invents (or is forced to invent) and embraces temporal process. Our thinking sets up an evolutionary sequence having both forward and backward movements. To enlarge a Kantian expression, the phenomena appear in their time-constrained reference frame as evolutionary sequences.

We are always at an intersection of the forward and backward movements,

T_1 and T_2 the turning-points of the transformations represented in the figure. We are always at the intersection of existence and reflection. That this becomes grasped, or simply that it is possible to grasp this, indicates a new intersection displaced beyond the whole set. This more enfolded intersection need not carry the same limit since the limit appears just at the point of inter-section where the whole pattern of existence/reflection becomes grasped or accepted as an essential operation of thought. The limit may well be tran-scended by further deepening of consciousness and extension of experience.

8. These arguments have dealt with some of the implications of the activity or movement of consciousness as it intelligently perceives and conceives order and harmony within the world. Consciousness is not derived from the trans-formations of the world, nor are the transformations of the world derived from the movement of consciousness. The transformations of the world and the movement of consciousness are manifestations of a deeper movement, or non-movement, being relatively autonomous. Consciousness is not identified with the material processes of thought, which are at least partially mechanical, but the multiple linkings or connections which exist are being examined. The foregoing sections sketch an investigation of the order of time which con-sciousness conceives in the form of an ordered set of transformations called evolution. The evolutionary perspective considered is nuanced in such a way that the theory of evolution is not wholly a creation of consciousness, nor is consciousness wholly embraced within a theory of evolution. One link between the two is expressed by the term 'involvement,' an example of which is reduction.

Through intelligent perception, consciousness grasps transformation as a central order of the world. We arrive at a causally ordered set of transfor-mations called evolution by tracing our links with the world within which we discover ourselves. We time-order some of our experiences beyond the immediate time of our existence, and I call this retrospective time-ordering reflection. I am not speaking of memory when I say that consciousness has a retrospective mode of activity; I am speaking of a time-ordering taking place under the direction of consciousness. Reflection is a characteristic of the time-constrained mode of conception; consciousness does reflect, in that retrospec-tive mode, upon its own tracings which extend into a remote past connected by the transformations or metamorphoses that are evolution.

The reciprocal linkings between man and the world, as suggested in the figure, do not imply that consciousness is confined to the world's transfor-mations. We are linked to the world by means of material conditions, and the figure includes our conscious awareness of some of those linkings. Some, because there are other linkings not implied by reflection, even in the material sense, and apart from these, our linkings with the world do not exhaust the being of consciousness. We might say the same thing in reverse from the point of view of the world: consciousness does not exhaust the being of the world. In a wider sense we can say also that neither the world nor humanity exhausts the deep ground from which both flow. To say that evolution (and I stress that I do not intend any goal or direction when I use this term) is a characteristic form or order of the world's becoming, or an order abstracted from the unknown becoming, is not to give a permanent essence to the world. Nor is it to give a permanent essence to consciousness to say that man understands that he is the world transformed, having arrived at that understanding through a movement that has the combined features of evolution and involvement or, similarly, the unified features of existence and reflection.

9. Both the theory of evolution and the quantum theory implicate man in the general transformations within this world, and this implication leads to deeper ideas of wholeness or non-separation. The limits I pointed to are relative limits appearing at intersections of evolution and involvement. They are not absolute limits in the sense of stopping or exhausting the activity of consciousness, but they point to the limitations of certain kinds of scientific process or of particular self-centred interpretations of science. The knowing which follows the newer types of wholeness connecting different orders will be quite different from the knowing which follows older patterns of separability in a polarized, fragmenting fashion. This may be a transformation of consciousness or the perception and conception of a deeper order and harmony.

Robert Rosen:
the Schrödinger question:
What is life? fifty years later

I. *General introduction*

Last year witnessed the 50th anniversary of the publication of Erwin
Schrödinger's essay *What Is Life?* It first appeared in print in 1944, based on a
series of public lectures delivered the preceding year in Dublin. Much has
happened, both in biology and in physics, during the half-century between
then and now. Hence, it might be appropriate to reappraise the status of
Schrödinger's question, from a contemporary perspective, at least as I see it
today. That is what we shall attempt herein.

I wonder how many people actually read this essay nowadays. I know I
have great difficulty in getting my students to read anything more than five
years old; that is their approximate threshold separating contemporary from
antiquarian, relevant from irrelevant. Of course, in the first decade or two of
its existence, as H.F. Judson (1979) says, 'everybody read Schrödinger,' and
its impact was wide indeed.

The very fact that 'everybody read Schrödinger' is itself unusual. For his
essay was a frank excursion into theoretical biology, and hence into something that most experimental biologists declare monumentally uninteresting
to them. Actually, I believe it was mostly read for reassurance. And, at least if
it is read superficially and selectively, the essay appears to provide that in
abundance; it is today regarded as an utterly benign pillar of current
orthodoxy.

But, as I will argue below, that is an illusion, an artifact of how
Schrödinger's exposition is crafted. I will argue that its true messages, subtly
understated as they are, are heterodox in the extreme, and always were. There
is no reassurance in them; indeed, they are quite incompatible with the dogmas of today. By the stringent standard raised by the Schrödinger question

'What is life?' following these dogmas has actually made it harder, rather than easier, to provide an adequate answer.

II. *What is life?*

Let us begin with the very question with which Schrödinger entitled his essay. Plainly, this is what he thought biology was *about*, what was its primary object of study. He thought that this 'life' was exemplified by or manifested in specific organisms, but that at root, biology was not about *them*; it concerned rather whatever it was about these particular material systems which distinguished them, and their behaviours, from inert matter.

The very form of the question connotes that Schrödinger believed 'life,' as such, is in itself a legitimate object of scientific scrutiny. It connotes a noun, not merely an adjective, just as, say, rigidity, or turbulence, or (as we shall get to later) openness does. Such properties are exemplified in the properties or behaviour of individual systems, but these are only *specimens*: the concepts themselves clearly have a far wider currency, not limited to any explicit list of such specimens. Indeed, we can ask a Schrödinger-type question, 'What is X?' about any of them.

I daresay that, expressed in such terms, the Schrödinger question would be dismissed out of hand by today's dogmatists as, at best, meaningless and, at worst, simply fatuous. It seems absurd in principle to partition a living organism, say a hippopotamus, or a chrysanthemum, or a paramecium, into a part which is its 'life' and another part which is 'everything else,' and even worse to claim that the 'life' part is essentially the same from one such organism to another, while only the 'everything else' will vary. In this view, it is simply outrageous to regard expressions like 'hippopotamus life' or 'chrysanthemum life' as meaningful at all, let alone equivalent to the usual expressions, 'living hippopotamus,' 'living chrysanthemum.' Yet it is precisely this interchange of noun and adjective which is tacit in Schrödinger's question.

This approach represents a turnabout which experimentalists do not like. On the one hand, they are perfectly willing to believe (quite deeply, in fact) in some notion of *surrogacy*, which allows them to extrapolate their data to specimens unobserved; to believe, say, that *their* membrane's properties are characteristic of membranes in general, or that the data from their rat can be extrapolated ad lib to other species (cf. Rosen 1983). On the other hand, they find it most disquieting when their systems are treated as the surrogatees, especially to be told something about *their* membrane by someone who has not looked at their membrane, but rather at what they regard as a physico-mathematical 'abstraction.' When pressed, experimentalists tend to devolve the notions of surrogacy they accept onto *evolution*: surrogates 'evolve' from each other, and hence, what does not evolve cannot be a surrogate. You cannot

have the issue both ways, and that is one of the primary Schrödinger unortho-
doxies, tacit in the very question itself.

A typical empiricist (not just a biologist) will tell you that the Schrödinger
question is a throwback to Platonic idealism, and hence completely outside
the pale of science. The question itself can thus only be entertained in some
vague metaphoric sense, regarded only as a *façon de parler* and not taken
seriously. But Schrödinger gives no indication that he intends only such metaphoric imagery; I think (and his own subsequent arguments unmistakably
indicate) that, to the contrary, he was perfectly serious. And Schrödinger
knew, if anyone did, the difference between Platonism and science.

III. *Schrödinger and 'new physics'*

Erwin Schrödinger was one of the outstanding theoretical physicists of our
century, perhaps of any century. He was a past master at all kinds of propa-
gation phenomena, of statistical mechanics and thermodynamics, and almost
every other facet of his field. Moreover, he viewed physics itself as the ultimate science of material nature, including, of course, those material systems
we call organisms. Yet one of the striking features of his essay is the constantly iterated apologies he makes, both for his physics and for himself personally. While repeatedly proclaiming the 'universality' of contemporary
physics, he equally repeatedly points out (quite rightly) the utter failure of its
laws to say anything significant about the biosphere and what is in it.

What he was trying to say was stated a little later, perhaps even more vividly, by Albert Einstein; in a letter to Leo Szilard, Einstein said: 'One can best
appreciate, from a study of living things, *how primitive physics still is.*'

Schrödinger (and Einstein) were not just being modest; they were pointing
to a conundrum, about contemporary physics itself, and about its relation to
life. Schrödinger's answer to this conundrum was simple, and explicit, and
repeated over and over in his essay. And it epitomized the heterodoxy I
have alluded to before. Namely, Schrödinger concluded that organisms
were repositories of what he called *new physics*. We shall turn a little later to
his gentle hints and allusions regarding what that 'new physics' would comprise.

Consider, by contrast, the words of Jacques Monod, writing some three
decades after the appearance of Schrödinger's essay: 'Biology is *marginal*
(my emphasis) because – the living world constituting but a tiny and very
"special" part of the universe – it does not seem likely that the study of living
things will ever uncover general laws applicable outside the biosphere.' With
these words, Monod opens his book *Chance and Necessity*, which sets out the
orthodox position. This idea of the 'marginality' of biology, expressed as a
denial of the possibility of learning anything new about matter (i.e., about

physics) by studying organisms, is in fact the very cornerstone of his entire development.

Monod did not dare to attack Schrödinger personally, but he freely condemned anyone else who suggested there might be 'new physics' wrapped up in organism, or in life, in the harshest possible way; he called them vitalists, outside the pale of science. Sydney Brenner, another postulant of contemporary orthodoxy, was even blunter, dismissing the possibility of a 'new physics' as 'this nonsense.'

But Schrödinger, within his own lifetime, had seen, and participated in, the creation of more 'new physics' than had occurred in almost the entire previous history of the subject. It did not frighten him; on the contrary, he found such possibilities thrilling and exhilarating; it was what he did physics for. Somehow, it is only the biologists it terrifies.

There is one more historical circumstance which should perhaps be mentioned here. Namely, biological thoughts were lurking very close to the surface in the cradles of the new quantum theory in the 1920s. Niels Bohr himself was always profoundly conscious of them. He had in fact grown up in an atmosphere of biology; his father was an eminent physiologist (the familiar Bohr effect, involving the cooperativity of binding of oxygen to hemoglobin, was named for Niels Bohr's father). Many of Bohr's philosophical writings, particularly those dealing with complementarity, are awash in biological currents. In general, the creators of the new quantum theory believed they had at last penetrated the innermost secrets of all matter. I have been told, by numerous participants and observers of these developments, of the pervasive expectation that the 'secrets of life' would imminently tumble forth as corollaries of this work.

That, of course, is not what happened. And indeed, Schrödinger's ideas about the 'new physics' to be learned from organisms lie in quite a different direction. We shall get to it presently.

IV. Genotypes and phenotypes

We have seen, in the preceding sections, just how radical and unorthodox Schrödinger's essay is, first in simply posing the question 'What is life?' and, second, in tying its answer to 'new physics.' Both are rejected, indeed condemned, by current dogmas, which cannot survive either of them. How, then, could this essay possibly have been read for reassurance by the orthodox?

The answer, as I have hinted above, lies in the way the essay is crafted. Viewed superficially, it looks primarily like an exposition of an earlier paper by Schrödinger's younger colleague, Max Delbrück. Delbrück was a student during the yeasty days in which the new quantum theory was being created, and was deeply impressed by the ambiences we have sketched above. Indeed, he turned to biology precisely because he was looking for the 'new physics' Schrödinger talked about, but he missed it. Delbrück's paper, on which Schrödinger dwelt at such length in his essay, argued that the 'Mendelian gene' had to be a molecule (but cf. Section VI below).

Today, of course, this identification is so utterly commonplace that no one even thinks about it any more – a deeply reassuring bastion of reductionism. But it is in fact much more complicated than it looks, both biologically and, above all, physically. As we shall see shortly, identifications require two different processes, and Delbrück only argued one. It was Schrödinger's attempt to go the other way, the hard way, roughly to deal with the question 'when is a molecule a Mendelian gene?' which led him to his 'new physics,' and hence to the very question 'What is life?'

At this point, it is convenient to pause to review the original notion of the Mendelian gene itself, a notion intimately tied to the genotype/phenotype dualism.

Phenotypes, of course, are what we can see directly about organisms. They are what behave, what have tangible, material properties we can measure and compare and experiment with. Gregor Mendel originally conceived the idea of trying to account for the similarities, and the differences, between the phenotypes of parents and offspring in a systematic way.

Mendel was, at heart, a good Newtonian. Newton's laws in mechanics say roughly that if *behaviours* are differing, then some *force* is acting. Indeed, that is how you always recognize a force, by the way it changes a behaviour; and that is how you measure that force. In these terms, Mendel's great innovation was to conceive of phenotype as *forced behaviours*, and to think of underlying 'hereditary factors' (later called genes) as forcers of these phenotypes. In a more philosophical parlance, his 'hereditary factors' constituted a new causal category for phenotypes and their behaviours; he was answering questions of the form 'why these phenotypic characters?' with answers of the form 'because these hereditary factors.'

As everyone knows, Mendel proceeded to measure the forcings of genotype by phenotype, by selecting a particular phenotype ('wild type') as a standard and comparing it to phenotypes differing from it in only 'one allele,' as we would now say.

Exactly the same kind of thing was then going on elsewhere in biology. For instance, Robert Koch was also comparing phenotypes and their behaviours; in this case, what he called 'healthy' (his analogue of 'wild type') and 'diseased.' The differences between them, the symptoms or syndromes marking the discrepancy between the former and the latter, were also regarded as forced, and the forcers called 'germs.' This, of course, constituted the 'germ theory' of disease.

To anticipate somewhat, we can see that any such 'genotype/phenotype' dualism is allied to the Newtonian dualism between states (or phases) and forces. The former are what behave; the latter are what make them behave. In a still earlier Aristotelian language, the states or phases represent material causation of behaviour; the forces are an amalgam of formal and efficient causation. In biology, the phenotypes are what get the states and behaviours; the genotypes or germs are identified as forces which drive them.

It is all too easy to simply posit forces in order to account for the tangible changes of behaviour which we can see directly. Critics of science have always pointed out that there is indeed something ad hoc, even ineluctably circular, in all this: to define a force in terms of observed behaviour, and then turn around and 'explain' the behaviour in terms of that posited force. Indeed, even many scientists regard the unbridled invention of such 'forces' as the entire province of 'theory,' and dismiss it accordingly, out of hand, as something unfalsifiable by observation of behaviour alone. Worst of all, perhaps, such a picture generally requires going 'outside' a system, to a larger system, to account for behaviours 'inside' it; this does not sit well with canons of reductionism, or with presumptions of 'objectivity' or 'context-independence' in which scientists like to believe. Finally, of course, we should not forget fiascos like phlogiston, the epicycles, the luminiferous ether, among many others, which were all characterized in precisely such a fashion.

For all these reasons, then, many people doubted the 'reality' of the Mendelian genes. Indeed, many eminently respectable physicists, for similar reasons, doubted the 'reality' of atoms until well into the present century (cf. the interesting discussion in Pais [1982]).

It is precisely at this point that the argument of Delbrück, which Schrödinger develops in such detail in his essay, enters the picture. For it proposes an identification of the functional Mendelian gene, defined entirely as a forcer of phenotype, with something more tangible; something with properties of its own, defined independently, a *molecule*. It proposes, as we shall see, a way to *realize* a force in terms of something more tangible which is generating it. But, as we shall now see, this involves a new, and perhaps worse, dualism of its own.

V. *On inertia and gravitation*

What we are at present driving towards is the duality between how a given material system changes its own behaviour in *response* to a force, and how that same system can *generate* forces which change the behaviour of other systems. It is precisely this duality which Schrödinger was addressing in the context of 'Mendelian genes' and 'molecules,' and the mode of forcing of phenotype by genotype. As we saw above, a relation between these two entirely different ways of characterizing a material system is essential if we are to remove the circularities inherent in either alone.

To fix ideas, let us consider the sardonic words of Ambrose Bierce, taken from his *Devil's Dictionary*, regarding one of the most deeply entrenched pillars of classical physics:

> GRAVITATION, n. The tendency of all bodies to approach one another, with a strength proportioned to the quantity of matter they contain – the quantity of matter they contain being ascertained by the strength of their tendency to approach one another. This is a lovely and edifying illustration of how science, having made A the proof of B, makes B the proof of A.

This, of course, is hardly fair. In fact, there are two quite different 'quantities of matter' involved, embodied in two distinct *parameters*. One of them is called *inertial mass*, and pertains to how a material particle *responds* to forces imposed on it. The other is called *gravitational mass*, and pertains rather to how the particle *generates* a force on other particles. From the beginning, Newton treated them quite differently, requiring separate *laws* for each aspect.

It so happens that, in this case, there is a close relation between the *values* of these two different parameters. In fact, they turn out to be numerically equal. This is in turn a most peculiar fact, one which was reviewed by Einstein not merely as a happy coincidence, but rather as one of the deepest things in all physics. It led Einstein to his 'Principle of equivalence' between inertia and gravitation, and this in turn provided an essential cornerstone of general relativity. But that is another story.

We clearly cannot hope in general for identical relations between 'inertial' and 'gravitational' aspects of a system such as are found in the very special realms of particle mechanics. Yet, in a sense, this is precisely what Schrödinger's essay is about. Delbrück, as we have seen, was seeking to literally reify a forcing (the Mendelian gene), something 'gravitational,' by clothing it in something with 'inertia,' by *realizing* it as a molecule. Schrödinger understood that this was not nearly enough, that we must also be able to go the other way and determine the forcings manifested by something characterized 'inertially.' In more direct language, just as we hope to realize a force by a thing, we must also, perhaps more importantly, be able to realize a thing by a force. It was precisely in this latter connection that Schrödinger put forward the most familiar parts of his essay: the 'aperiodic solid,' the 'principle of order from order,' and the 'feeding on negative entropy.' And as suggested earlier, it was precisely here that he was looking for the 'new physics.' We shall get to all this shortly.

Before doing so, however, we must look more closely at what this peculiar dualism between the 'inertial' and the 'gravitational' aspects of a material system actually connotes.

Newton himself was never much interested in understanding what a force *was*; he boasted that he never even asked this question. That was what he meant when he said: 'Hypothesis non Fingo.' He was entirely interested in descriptions of system behaviours, which were rooted in a canonical state space or phase space belonging to the system. Whatever force 'really' was, it was enough for Newton that it manifested itself as a *function* of phase, i.e., a function of something already inside the system. And that is true, even when the force itself is coming from *outside*.

This, it must be carefully noted, is quite different from *realizing* such a force with an 'inertia' of its own, generally quite unrelated to the states or phases of the system being forced. This latter, as we have seen, is what Schrödinger and Delbrück were talking about, in the context of the 'Mendelian gene' as a forcer of phenotype. Newton himself, as we have seen, did not care much about such realization problems; consequently, neither did the 'old physics' which continues to bear his personality. Indeed, this is perhaps the primary reason that Schrödinger, who increasingly saw 'life' as wrapped up precisely with such realization problems, found himself talking about 'new physics.' It is precisely the tension between these two pictures of force which will, one way or another, dominate the remainder of our discussion.

We must next call attention to the central role played in the original Newtonian picture by the *parameters* he introduced, exemplified by 'inertial mass' and 'gravitational mass.' Roughly, these serve to couple states or phases (i.e., whatever is behaving) to forces. In mechanics, these parameters are independent of both phases and of forces, independent of the behaviours they modulate. Indeed, there is nothing in the universe which can change them, or touch them in any way. Stated another way, these parameters are the quintessence of objectivity, independent of any context whatever.

Further, if we are given a Newtonian particle, and ask ourselves what *kind* of particle, what 'species' of particle it is, the answer lies not in any particular *behaviour* it manifests under the influence of one or another force impressed on it, not in the states or phases which do the behaving, but rather, precisely in those parameter values, its masses. They are what determine the particle's *identity*, and in this sense, *they are its genome*; the particular behaviours the particle may manifest (i.e., how its phases or states are changing when a force is imposed on it) are accordingly only *phenotypes*. Nor does this identity reside in the behaviours of other systems, forced by it.

In causal language, the parameters of which we are speaking constitute *formal cause* of the system's behaviours or phenotypes (the states themselves are their material causes; the forces are efficient causes).

Thus, there is a form of the phenotype/genotype dualism arising already here, where *genome* (in the sense of 'species-determining,' or 'identity-determining') is associated with *formal causes* of behaviours or phenotypes. It arises here as a consequence of the dualism mentioned earlier, between the states or phases of a system and the forces which are making it behave.

We invite the reader to ponder these last remarks, in the context of the realization problems which Schrödinger (and to a much lesser extent, Delbrück) were addressing. You will begin to see, I believe, that it is not quite as straightforward as current dogmas would indicate. We will return to these matters shortly.

VI. *'Order from order'*

We will digress from the conceptual matters we have been considering and look briefly at Schrödinger's essay into the realization problems we discussed earlier. In general, he was concerned with turning inertia into gravitation, a thing into a force, a molecule into a 'Mendelian gene.' This is perhaps the most radical part of Schrödinger's argument, which ironically, is today perceived as an epitome of orthodoxy.

As noted earlier, Delbrück had argued that the Mendelian gene, as a forcer of phenotype, must be inertially realized as a molecule. The argument was as follows: whatever these 'genes' *are*, in material terms they must be small. But small things are, by that very fact, generally vulnerable to thermal noise. Genes, however, must be stable to (thermal) noise. Molecules are small and stable to thermal noise. *Ergo*, genes must *be* molecules. Not a very cogent argument, perhaps, but the conclusion was satisfying in many ways; it had the advantage of being *anschaulich*. Actually, Delbrück's arguments only argue for *constraints*, and not just holonomic, Tinkertoy ones like rigidity; the same arguments are just as consistent with, for example, two molecules per 'gene,' or three molecules, or *N* molecules, or even a fractional part of a molecule.

Schrödinger was one of the first to tacitly identify such constraints with the concept of *order*. Historically, the term 'order' did not enter the lexicon of physics until the latter part of the 19th century, and then only through an identification of its negation, *disorder*, with the thermodynamic notion of entropy. That is, something was ordered if it was not disordered, just as something is non-linear if it is not linear.

As we have discussed at length elsewhere, constraints in mechanics are identical relations among state or phase variables and their rates of change. If configurational variables alone are involved, the corresponding constraint is called *holonomic*. Rigidity is a holonomic constraint. The identical relations making up the constraint allow us to express some of the state variables as functions of the others, so that not all the values of the state variables may be freely chosen. Thus, for example, a normal chunk of rigid bulk matter, which from a classical microscopic viewpoint may contain 10^{30} particles, and hence three times that number of configurational variables, can be completely described by only six. Such heavily constrained systems are often referred to nowadays as *synergetic* (cf., e.g., Haken [1977]; he calls the independently choosable ones 'controls,' and the remaining ones 'slaved'). We might note, in passing, that traditional bifurcation theory is the mathematics of breaking constraints; its classical problems, like the buckling of beams, and other failures of mechanical structures, involve precisely the breaking of rigid constraints, as a function of changing parameters associated with impressed *forcings*. The reader should bear this in mind in the light of the discussion of the preceding section.

Non-holonomic constraints, which involve both configuration variables and their rates of change, have received much less study, mainly because they are not mathematically tidy. But they are of the essence to our present discussion, as we shall see.

The language of constraints as manifestations of 'order' can be made compatible with the language of entropy coming from thermodynamics, but the two are by no means equivalent. Schrödinger took great pains to distinguish them, associating the latter with the 'old physics,' and embodied in what he called 'order from disorder' marking a transition to equilibrium in a closed system. But by speaking of 'order' in terms of constraints, he opened a door to radically new possibilities.

Schrödinger obviously viewed phenotypes, and their behaviours, as *orderly*. At the very least, the behaviours they manifest, and the rates at which these behaviours unfold, are obviously highly constrained. In these terms, the constraints involved in that orderliness are inherently non-holonomic, viewed from the standpoint of phenotype alone.

As we have seen, the Mendelian gene was introduced as a 'forcer' of phenotype. Delbrück had argued that such a Mendelian gene was, in material ('inertial') terms, a molecule, mainly on the grounds that molecules were rigid. Thus, whatever 'order' there is in a molecule entirely resides in its constraints. But these, in turn, are *holonomic*. As Schrödinger so clearly perceived, the real problem was to somehow move this *holonomic* order, characteristic of a molecule, into the *non-holonomic* order manifested by a phenotype (which is not a molecule). In the more general terms we have outlined in the preceding section, the problem is to realize an 'inertial,' structural, holonomic thing in terms of a force exerted on a dynamic, non-holonomic thing.

This was the genesis of Schrödinger's conception of 'order from order.' Or, more precisely, large-scale, non-holonomic, phenotypic order, being forced by small-scale, rigid, holonomic, molecular order. It was this kind of situation for which Schrödinger found no precedent in the 'old physics.' This was why, in his eyes, organisms resisted the 'old physics' so mightily.

As everyone knows, Schrödinger expressed the holonomic order he perceived at the genetic end in the form of the 'aperiodic solid.' In other words, not just *any* holonomic or rigid structure could inertially realize a 'Mendelian gene,' but only certain ones, which both specialized and generalized conventional molecules in different ways. Nowadays, it is axiomatic to simply identify 'copolymer,' and indeed, with DNA or RNA, and the constraints embodying the holonomic order with 'sequence.' But this changing of names, even if it is justified (and I claim it is not), does not even begin to address the realization problem, the transduction of genomic 'inertia' into 'gravitation' which Schrödinger was talking about.

Schrödinger was perhaps the first to talk about this transduction in a cryptographic language – to express the relation between holonomic order in genome, and non-holonomic order in phenotype, as constituting a *code*. This view was seized upon by another physicist, George Gamow, a decade later; after contemplating the then-new Watson-Crick structure for DNA, he proposed a way to use DNA as a template, for moving its holonomically constrained 'order' up to another holonomically constrained but much less rigid inertial thing, protein. This is a very far cry from the 'code' which Schrödinger was talking about; it is at best only an incremental syntactic step. The next big one would be to solve the 'protein-folding problem,' something over which the 'old physics' claims absolute authority. After three decades of fruitless, frustrating, and costly failures, the field is just beginning to move again: ironically, by postulating that protein folding is a forced rather than spontaneous process; trying to realize these putative forcers in 'inertial' terms, and thus in a sense replaying the Mendelian experience in a microcosm. But this again is another story.

In addition to the principle of 'order from order' which Schrödinger introduced to get from genotype to phenotype, and the 'aperiodic solid' which he viewed as constituting the 'genetic' end of the process, and the idea of a cryptographic relation between holonomic constraints in genotype and the non-holonomic ones characterizing phenotype, Schrödinger introduced one more essential feature. That was the idea of *feeding* (on 'negative entropy,' he said, but for our purposes it does not matter what we call the food). This was not just a gratuitous observation on Schrödinger's part. He was saying that, in order that the entire process of 'order from order' work at all, the system exhibiting it *has to be open* in some crucial sense. In the next section, we shall look at this basic conclusion in more detail.

As we stated at the outset, Schrödinger's essay was read by people, particularly molecular biologists, for reassurance. The reassurance lay mainly in Schrödinger's use of innocent-sounding terms in familiar contexts. But, as I hope is already becoming clear, whatever this essay may offer, reassurance is not present there.

VII. *The 'open system'*

From the foregoing, we see that Schrödinger envisioned two entirely different ways in which biological phenotypes, considered as material systems, are open. On the one hand, they are open to forcings, embodied tacitly in the Mendelian genes. On the other hand, they are also open to what they feed on, what they 'metabolize.' The former involves the effects of something *on* phenotype; the latter involves the effects of phenotype on something else (specifically, on 'metabolites' residing in the environment). Schrödinger was tacitly suggesting a profound connection between these two types of openness: namely, that a system open in the first sense must also be open in the second. Or, stated another way, that the entire process of 'order from order' which he envisioned, and indeed, the entire Mendelian process which it represented, cannot work in a (thermodynamically) closed system at all.

Such thermodynamically open systems accordingly can be considered as 'phenotypes without genotypes.' They are the kinds of things which Mendelian genes can force. So this is a good place to start; especially since, as we shall see, it is already full of 'new physics,' even without any explicit genome to force it. To anticipate somewhat, we will be driving toward a new perspective on Schrödinger's inverse question 'when can a molecule be a Mendelian gene?' in terms of another question, of the form 'when can a thermodynamically open system admit Mendelian forcings?'

The history of ideas pertaining to 'open systems' is in itself interesting, and merits a short statement. The impetus to study them, and their properties, came entirely from biology, not at all from physics, which preferred to rest content with its closed, isolated, conservative systems and their equilibria, and to blithely assign their properties a universal validity.

The first person to challenge this prevailing attitude, to my knowledge, was Ludwig von Bertalanffy in the late 1920s. Ironically, he was attempting to combat the frank vitalism of the embryologist Driesch, particularly in regard to embryological or developmental processes then given the name 'equifinality.' Bertalanffy showed that these phenomena, which so puzzled Driesch, simply evaporated once we gave up the strictures of thermodynamic closure, and replace the concept of equilibrium by the far more general notion of steady state (*Fliessgleichgewicht*), or ultimately, the still more general types of attractors which can exist in open systems.

It is no accident that Bertalanffy was a person whom Jacques Monod (cf. above) loathed, and whom he (among many others) castigated as a 'holist.' Obviously, by their very nature, open systems require going outside a system, going from a smaller system to a larger one, to understand its behaviours. Stated another way, openness means that even a complete understanding of internal 'parts' or subsystems cannot, of itself, account for what happens when a system is open. This in turn flies in the face of the 'analysis' or reductionism which Monod identified with 'objective science.' But again, this is another story.

In the late 1930s, Nicolas Rashevsky discovered some of the things that can happen in a specific class of such open systems, at present termed reaction-diffusion systems. He showed explicitly how such systems could spontaneously establish concentration gradients in the large. This is, of course, the most elementary morphogenetic process and, at the same time, is absolutely forbidden in thermodynamically closed systems. It might be noted that another name for this process, in physiology, is 'active transport.' Over a decade later, this process was rediscovered by Alan Turing, in a much simpler mathematical context than Rashevsky had used. A decade after that, the same phenomena were picturesquely characterized by Ilya Prigogine, a chemical thermodynamicist, under the rubric of 'symmetry breaking.' A huge literature on pattern generation, and 'self-organization' in general, has arisen in the meantime, based on these ideas.

Bertalanffy himself was quite well aware of the revolution in physics which was entailed in his concept of the 'open system.' Indeed, he said quite bluntly: 'The theory of open systems has opened up an entirely *new field of physics.*' Quite early in the game (in 1947), Prigogine likewise said: 'Thermodynamics is an admirable but *fragmentary* theory, and this fragmentary character originates from the fact that it is applicable only to states of equilibrium in closed systems. Therefore, it is necessary to establish a broader theory ...'

Parenthetically, I would assert that, even today, there is as yet no acceptable *physics* of open systems. This is because 'closed systems' are so degenerate, so non-generic, that when you open them, the resultant behaviour depends on how they were opened much more than on what they were like when closed. This is true even for the classical theory of thermodynamics itself, and why this classical theory does not lend itself to expansion into a true physical theory of open systems. What passes for theory at this level is entirely phenomenological, and is expressed in dynamic language, not thermodynamic. These facts, it should be noted, are of direct and urgent concern to experimental analysis, particularly in biology, since the very first step in any analytic procedure is to open the system up still further, in a way which is itself not reversible. That is, roughly, why 'analysis' and 'synthesis' are not in general inverse processes (cf. Section X(f) below).

In any case, Schrödinger himself could have known about these incipient revolutions in the 'old physics,' tacit in systems which feed and metabolize. But he had fixed his attention entirely on 'molecules,' and on biochemistry, and hence he missed a prime example of the very thing he was asserting, and which most biologists were even then denying; namely, that organisms teach new lessons about matter in general.

Open systems thus constitute in themselves a profound and breathtaking generalization of 'old physics,' based as it is on the assumption of excessively restrictive closure conditions, conservation laws, and similar non-generic presumptions which simply do not hold for living things. Seen in this light, then, is it really biology which is, in Monod's words, 'marginal,' 'a tiny and very special part of the universe,' or is it rather the 'old physics'? In 1944, Schrödinger suggested that it was the latter which might be the case. Today, fifty years later, that possibility continues to beckon, and indeed, with ever-increasing urgency.

VIII. *The forcing of open systems*

The behaviours manifested in open systems, such as their capacity to generate and maintain stable spatial patterns, exemplify neither the classical thermody-namic notion of 'order from disorder,' as Schrödinger used the term, nor what he called 'order from order.' As we have said, open system behaviours look like phenotypes, but they are not 'forced,' in any conventional sense; certainly not in any Mendelian sense, event though they have 'genomes' expressed in their parameters. Nevertheless, their behaviours can be stable without being rigid, or in any sense holonomically constrained. Let us see what happens when we impose forcings on such a system and, especially, when we try to 'internalize' those forcings.

The essence of an 'open system' is, as we have seen, the necessity to invoke an 'outside,' or an environment, in order to understand what is going on 'inside.' That is, we must go to a larger system, and not to smaller ones, to account for what an open system is doing. That is why reductionism, or analy-sis, which only permits us to devolve system behaviour upon subsystem behaviours, fails for open systems. And as we have seen, that is why there is so much 'new physics' inherent in open systems. That fact, of course, does not make openness unphysical; it simply points up a discrepancy between the physics we now know and the physics we need.

But there are many ways a system can be open. So far, we have discussed only thermodynamic openness, characterized by energetic and material fluxes through the system. These are characterized by corresponding sources and sinks generally residing outside the system itself, in its environment. As we have seen, inherent in this view is the notion of the system exerting forces on its environment, acting as a 'pump' and driving the flow from sources to sinks.

But an open system in this thermodynamic sense can itself be forced; i.e., the environment can itself impress forces on the system. This is what we called a 'gravitational' effect earlier, and is in general a quite different kind of openness to environmental influence than the thermodynamic openness we have just been considering. System behaviour under the influence of such impressed forces has always been, of course, the lifeblood of classical particle mechanics, and also, in a somewhat modified form, of what is today roughly called control theory.

If there is already much 'new physics' in the free behaviours of open systems, we should not be surprised to find much more in their forced behaviours. Especially so since our intuitions about how material systems respond to impressed forces are generally drawn from very simple systems; indeed, generally linear ones. One of these intuitions, embodied in such things as servomechanisms and homeostats, is that a forced system will generally end up tracking the forcing. If this is so, it is correct to say that the relation between such an impressed force and the resulting system behaviour is ultimately a cryptographic one; the explicit relation between the two is embodied in the familiar *transfer function* of the system. That is already suggestive, but it is very risky to simply extrapolate such ideas to open systems.

As we have already emphasized, a system which is open in *any* sense is one whose behaviours depend on something outside the system itself; in a closed system, there *is* no outside. Thus, it has always been a tempting idea to 'internalize' the external influences in some way, to get a bigger system which is closed, and deal with that. Unfortunately, the genericity of openness forbids it; genericity in this sense means that openness is preserved under such perturbations. Indeed, what you end up with in this fashion is generally a bigger *open* system, which is in some sense even 'more open' than the one you started with. This is, in itself, an important observation, which among other things underlies the familiar notion of the 'side-effect,' but that again is another story. At any rate, what you typically end up with in carrying out such a strategy is the entire universe, which is not very helpful.

In general, the unforced or free situation in any system is one in which every force in the system is an internal force. In the language we introduced earlier, it is a situation in which every 'gravitational' aspect in the system can be assigned to a corresponding 'inertial' aspect of that system. But if a force is impressed on such a system from outside, that force has no 'inertial' correlate within the system; there is in some an excess of gravitation over available inertia, an 'inertial defect,' if you will.

Thus, if we wish to try to internalize such a force, we must augment our original system with 'more inertia'; in practice, that means adding more state variables and more parameters to the system in such a way that the forced behaviour of the original system is now free behaviour of the larger system.

Now, as we noted earlier, the effect of any force is to modify a rate from what it would be in an unforced or free situation. That is, a force shows up in the system as an acceleration or deceleration of some system behaviour, i.e., as a *catalyst*. If we can internalize such a force in the manner we have described, in terms of 'inertially' augmenting the original system with more state variables and more parameters, then it is not too much an abuse of language to call the new variables we have introduced (and of course the parameters we need to couple them to the original system) *enzyme.*

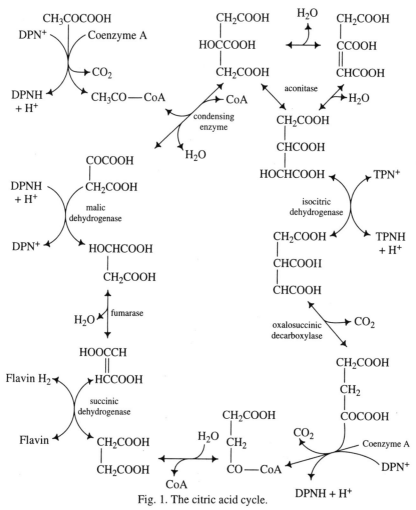

Fig. 1. The citric acid cycle.

In formal terms, such augmented systems must be very heavily constrained, with all kinds of identical relations between the new variables and parameters we have added (i.e., the 'enzymes') and the tangent vectors which govern change of state in the system. That is, the new variables are doing a 'double duty': they define state in the larger system and also participate in operating on that state, i.e., in determining the rate at which such a state is changing.

Without going into details, these constraints are strong enough to be expressed in an abstract graphical language. A primitive example of this is the familiar representations of intermediary metabolism, such as that displayed in Figure 1. Here, the arrows represent the 'enzymes,' the inertial variables and parameters we have added to internalize impressed forces, while the vertices roughly correspond to state variables of the smaller open system on which the forcings are impressed. The graph thus expresses exactly the constraints we have just spoken of. The existence of such a graph is in fact a corollary of internalizing forces impressed on open systems in the manner we have discussed; not only in biology, but quite generally. To a large extent, the converse is also true, but that is not of immediate concern. We cannot help noting that such a graph looks very much like an 'aperiodic solid,' and indeed, a little reflection will reveal that it possesses many of the properties Schrödinger ascribed to that concept. The novel thing is that it is not a 'real' solid. It is, rather, a pattern of causal organization; it is a prototype of a *relational model* (cf. Rosen 1991).

Since the larger system is itself open, the new variables and parameters we have added to internalize impressed forces (i.e., the 'enzymes') will themselves have sources and sinks. They are not present in the above diagram, but without them, the enlarged system, represented by the graph, is generally *not stable* as a free system. If we want it to be stable, *we need more forces* impressed on the system to stabilize it. This is, roughly, where the Mendelian genes enter the picture.

In a nutshell, stabilization of this kind is attained by modulating the rates which the 'enzymes' impose on the original open system with which we started. This, in fact, is precisely what the Mendelian genes do; they correspond to accelerations or decelerations of the rates at which 'enzymes' themselves control rates. We may further think to 'internalize' impressed forces of this kind in exactly the same way we just internalized the 'enzymes' themselves: namely, add still more 'inertial' variables of state, and still more parameters to couple them to what we already have, to obtain an even bigger open system, and one which is even more heavily constrained than before. Just as before, these constraints are strong enough to be expressed in graphical language, but the kind of graph which arises at this level is much more complicated than heretofore. Roughly, instead of two levels of 'function,' embodied in the distinction we have drawn between the arrows of the graph and its vertices, we now have three such levels (roughly, the original metabolites, the 'enzymes' which force them, and now the Mendelian genes which force the 'enzymes'). If the original graphical structures are indeed thought of as 'aperiodic solids,' so too are the new ones, albeit of quite a novel type.

Unfortunately, even thus augmented, the resulting open systems are still not in general stable. The first thought here is to iterate yet again the process we have already used: namely, posit new impressed forces to modulate the Mendelian genes we have just internalized, seek to internalize them by means of still more 'inertia' (i.e., more state variables, more parameters to couple them to what is already in the system, and more constraints imposed upon them). But at this point, one can already glimpse an incipient regress establishing itself.

The only alternative to this infinite regress is to allow the sources and sinks for the internalized inertial forcers introduced at the Nth stage of such a process to have already arisen at earlier stages. A source for such an Nth-stage internalized forcer is *a mechanism for its replication*, expressed in terms of the preceding $N - 1$ stages, and not requiring a new $(N + 1)$th stage. Thus, replication is not just a formal means of breaking off a devastating infinite regress, but it serves precisely to stabilize the open system we arrived at in the Nth stage.

In biology, N seems to be a small number – 2 or 3, or perhaps 4 in multicellulars. But I can see no reason why this should be so in general.

Breaking off such an infinite regress does not come for free. In order for it to happen, the graphs to which we have drawn attention, and which arise in successively more complicated forms at each step of the process, must fold back on each other in unprecedented ways. In the process, we create (among other things) closed loops of efficient causation. As we have explained at great length elsewhere (cf. Rosen 1991) systems of this type cannot be simulated by finite-state machines (e.g., Turing machines); hence they themselves are not machines or mechanisms. In formal terms, they manifest impredicative loops. Systems of this type are what I call *complex*; among other things, they possess no largest (simulable) model. The physics of such complex systems, which we have described here in terms of the forcing of open systems (and they can be approached in many other ways; cf. Rosen op. cit.), is, I assert, some of the 'new physics' for which Schrödinger was looking.

IX. *When is a molecule a Mendelian gene?*

As we have seen above, this was the real question Schrödinger was addressing in his essay, the inverse of the question Delbrück thought he answered by asserting that a gene is a molecule.

The question looks intriguing because, at its root, it embodies a correspondence principle between an 'inertial' thing (e.g., a molecule), and a 'gravitational' thing (a force imposed on an open system). But as the discussion of the preceding section makes clear, the question is much more context-dependent than that; its answer involves not just inherent properties of a 'molecule' in itself (e.g., 'aperiodicity'), but also the properties of what system is being forced, and the preceding levels of forcing of which 'genome' is to be the last.

Thus, very little remains of Schrödinger's simple cryptographic picture of 'order from order,' in which rigid molecular structures get transduced somehow into non-rigid phenotypic ones. Rather, the initial 'order' appears as a pattern, or graph, of interpenetrating constraints, which determines what happens, and how fast it happens, and in what order, in an underlying open system. The arrows in these graphs, which we suggest constitute the real 'aperiodic solid,' are operators; they express 'gravitational' effects on the underlying system. To speak of them in terms of 'inertia,' it is much more appropriate to speak of active sites than of molecules. The two are not the same.

Indeed, at this point, much simpler questions – e.g., 'when is a molecule an enzyme?' – are hard to approach in purely inertial terms. These are all 'structure-function' questions; they are all hard because a 'function' requires an external context whereas a 'structure' does not.

In a certain sense, if all you want to talk about is an 'active site' (i.e., something gravitational), and you find yourself talking about a whole molecule (i.e., something inertial), you run a severe risk of losing the site in the structure. There is, in a sense, much more inertia in a whole molecule than in a functional 'site.' We spoke earlier of impressed forces, imposed from the environment of a system, constituting an 'inertial defect'; structure-function problems tend to involve a dual 'inertial excess' of irrelevant information.

There is some 'new physics' here too, I would wager.

X. *What is life?*

In this penultimate section, we shall review the Schrödinger question in the light of the preceding discussions, and in terms of a number of subsidiary questions either raised directly by Schrödinger himself, or which have come up along the way.

(a) *Is 'What is life?' a fair scientific question?* My answer is 'of course, it is.' Not only is it a fair question, it is ultimately what biology is about; it is the central question of biology. The question itself may be endlessly rephrased, but to try to eliminate it in the name of some preconceived ideal of mechanistic 'objectivity' is in itself a far more subjective thing to do than that ideal itself allows.

(b) *Does the answer involve 'new physics'?* Once you admit questions of the Schrödinger type, which treats an adjective or predicate as a thing in itself, you are already doing 'new physics.' More formally, the 'old physics' rests on a dualism between phases or states, and forces which change the states, which make the system 'behave.' Predicates, or adjectives, typically pertain to these behaviours, which are what we see directly. Moreover, the emphasis here is overwhelmingly skewed in the direction of what I have called above the 'inertial' aspects of a system, how it responds to forces, at the expense of its 'gravitational' aspects, or how it itself exerts forces. In biology, this shows up in terms of 'structure-function problems,' where 'structure' pertains to 'inertia' and 'function' to 'gravitation.'

Many biologists, indeed the same ones who would deny the legitimacy of the Schrödinger question, assert that 'function' is itself an unscientific concept; in effect, they assert there is only 'structure.' Hence, biology can be scientific only insofar as it succeeds in expressing the former in terms of the latter. That is why Delbrück's argument, that a functionally defined 'Mendelian gene' comprises a familiar chemical structure, a molecule, was received so enthusiastically, while the converse question (roughly, when can a 'molecule' manifest such a function?), with which Schrödinger's essay is really concerned, was not even perceived.

Schrödinger's 'new physics,' embodied generally in his initial question, and specifically in his appraisal of the relation between genes and molecules, rests in his turning our inertial biases upside down, or at least suggesting that 'inertial' and 'gravitational' aspects of material systems be granted equal weight. Once this is done, 'new physics' appears of itself.

(c) *Is biology 'marginal'?* As we saw above, Jacques Monod used this word in expressing his claim that organisms are nothing but specializations of what is already on the shelf provided by 'old physics,' and that to claim otherwise was mere vitalism. He buttressed this assertion by observing that organisms are in some sense rare, that most material systems are not organisms.

This kind of argument rests on a confusion, or equivocation, concerning the term 'rare,' and identifying it with 'special.' An analogous argument could have been made in a humble area like arithmetic, at a time when 'most' numbers of ordinary experience were rational numbers, the ratios of integers. Suddenly a number like π shows up, which is not rational. It is clearly rare, in the context of the rational numbers we think we know. But there is an enormous world of 'new arithmetic' locked up in π, arising from the fact that it is much too general to be rational. This greater generality does not mean that there is anything 'vitalistic' about π, or even anything unarithmetic about it; indeed, the only 'vitalistic' aspects show up in the mistaken belief that number means rational number.

Schrödinger's 'new physics' makes an analogous case that organisms are more general than the non-organisms comprehended in the 'old physics,' and that their apparent rarity is only an artifact of sampling.

(d) *What is this 'new physics'?* Roughly, the 'new physics' involves going from special to general, rather than the other way around. At the very least, it means going from closed systems to open ones, discarding specializing hypotheses like closure conditions and conservation laws. As noted earlier, there is still no real 'physics' of such open systems, largely because the formalisms inherited from the 'old physics' are still much too special to accommodate it.

Most significant, I feel, will be the shifting of attention from exclusively 'inertial' or structural concepts to 'gravitational' aspects. This can be expressed as a shift from concerns with material causations of behaviour, manifested in state sets, to formal and efficient causations. As we have suggested above, these are manifested in graphical structures, whose patterns can be divorced entirely from the state sets on which they act. The mathematical precedent here lies in geometry, in the relation between groups of transformations tied to an underlying space, and the abstract group which remains when that underlying space is forgotten. To a geometer, concerned precisely with a particular space, this discarding of the space seems monstrous, since it is his very object of study; but to an algegraist, it throws an entirely new perspective on geometry itself, since the same abstract group can be *represented* as a transformation group in many different ways (i.e., an underlying space restored, which can look very different from the original one from which the group was abstracted). In the same way, it would look monstrous to a biologist, say, to throw away his state spaces (his category of material causation, his 'inertia') and retain an abstract graphical pattern of formal and efficient causation, but that is what is tacit in Schrödinger's concern with 'gravitation.'

(e) *What is life?* The lines of thought initiated in Schrödinger's essay lead inexorably to the idea that 'life' is concerned with the graphical patterns we have discussed, however briefly and inadequately, in the above discussion.

The formal metaphor we have suggested above – namely, dissociating a group of transformations from a space on which it acts – shows explicitly a situation in which what is a predicate or an adjective from the standpoint of the space can itself be regarded as a thing (the abstract group) for which an underlying space provides predicates. This is exactly analogous to the inversion of adjective and noun implicit in Schrödinger's question itself; as we saw at the outset, it involves partitioning an organism into a part which is its 'life' and a part which is 'everything else.' Seen from this perspective, the 'life' appears as an invariant graphical pattern of formal and efficient causation, as a 'gravitational' thing, and the 'everything else' in the form of material causation (e.g., state sets) on which such a graph can operate.

We have argued above that such a system must be *complex*. In particular, it means that such a system must have non-simulable models; it cannot be described as software to a finite-state machine. Therefore, it itself is not such a machine. There is a great deal of 'new physics' involved in this assertion as well.

To be sure, what we have been describing are necessary conditions, not sufficient ones, for a material system to be an organism. That is, they really pertain to what is not an organism, i.e., to what life is not. Sufficient conditions are harder; indeed, perhaps there aren't any. If not, biology itself is more comprehensive that we at present know.

(f) *What about 'artificial life'?* The possibility of 'artificial' or 'synthetic' life is certainly left wide open by the above discussion. However, the context it provides certainly excludes most, if not all, of what is at present offered under this rubric.

The first point to note is that, in open systems generally, analysis and synthesis are not inverse operations. Indeed, most analytic procedures do not even have inverses, even when it comes to simple systems or mechanisms. For instance, we cannot solve even an N-body problem by 'reducing' it to a family of $(N - k)$-body problems, whatever k is. How much more is this true in the kinds of material systems we have called complex, which we have argued is a necessary condition for life? Indeed, no one has ever really studied the problem of when an analytic mode possesses an inverse; i.e., when an analytic mode can be run backward, in any physical generality.

A second point is that what is currently called 'artificial life,' or 'A-life,' primarily consists of exercises in what used to be called biomimesis. This is an ancient activity, based on the idea that if a material system exhibits 'enough' of the behaviours we see in organisms, it must *be* an organism. Exactly the same kind of inductive inference is seen in the 'Turing Test' in 'artificial intelligence': a device exhibiting 'enough' properties of intelligence *is* intelligent.

In the preceding century, biomimesis was mainly pursued in physical and chemical systems; e.g., mimicking phenomena like motility, irritability, and tropisms, in droplets of oils embedded in ionic baths. Previously, it was manifested in building clockworks and other mechanical automata. Today, the digital computer, rather than the analog devices previously employed, provides the instrument of choice, a finite-state machine.

At root, these ideas are based on the supposition that some finite number (i.e., 'enough') of simulable behaviours can be pasted together to obtain something alive. Thus, that organisms are themselves simulable as material systems, and hence are not complex in our sense. This is a form of what is called Church's thesis, which imposes simulability as, in effect, a law of physics, and indeed, one much more stringent than any other. Such ideas already fail in arithmetic, where what can be executed by a finite-state machine (i.e., in an 'artificial arithmetic'), or in any finite (or even countably infinite) collection of such machines, is still infinitely feeble compared to 'real' arithmetic itself (this is Gödel's theorem).

In this connection, it might be observed that Schrödinger himself, in the last few pages of his essay, quite discounted the identification of organism with 'machine.' He did this essentially on the grounds that the latter are rigid, essentially low-temperature objects, while phenotypes are not. This provocative assertion, more-or-less a small aside remark on Schrödinger's part, is well worth pursuing in the context of what we have just said about the material basis of 'artificial life.'

XI. *Conclusions*

From this discussion, we can clearly see, as we said at the outset, that Schrödinger's essay, published a half-century ago, provides little comfort to an exclusively empirical view of biology; certainly not insofar as the basic question 'What is life?' is concerned. On the contrary: it removes the question from the empirical arena entirely, and in the process raises troubling questions, not only about biology, but about the nature of the scientific enterprise itself. But Schrödinger also proposed directions along which progress can be made. The consignment of Schrödinger's essay to the realm of archive is premature; indeed, it is again time that 'everybody read Schrödinger.'

REFERENCES

Haken, H. 1977. *Synergetics*. Berlin: Springer-Verlag
Judson, H.F. 1979. *The Eighth Day of Creation*. New York: Simon & Schuster
Pais, A. 1982. *Subtle Is the Lord*. Oxford University Press
Rosen, R. 1983. *American Journal of Physiology*. 244: 591–599
– 1991. *Life Itself*. New York: Columbia University Press
Schrödinger, E. 1944. *What Is Life?* Cambridge University Press

Appendix:
the troubles of quantum theory

THE COPENHAGEN INTERPRETATION

Most of us have no problem discussing the world in a rational way. We experience no dislocation between a rational description and one which appeals to our intuition. In thinking about everyday objects we seldom run into paradox or confusion. But this is not true when we confront the world of atoms, and this difficulty lies at the heart of many of the problems discussed in this book.

Werner Heisenberg was the discoverer of a formalism – quantum mechanics – which enabled scientists to describe mathematically the results of their experiments on atoms and elementary particles. However, when scientists began to argue about the meaning of this theory they found that paradoxes arose in their discussions. For example, Heisenberg's uncertainty principle showed that there was a degree of ambiguity in describing the properties of quantum objects and that these 'properties' depended upon the way and even the order in which they were measured. It seemed that the common-sense idea of 'properties possessed by an object' was not really appropriate in the quantum world.

In an effort to give quantum mechanics a rational foundation Heisenberg and Niels Bohr held a series of discussions at Copenhagen on how scientists should talk about quantum theory and its results. Bohr has written a clear exposition of what has become known as the Copenhagen interpretation but his arguments are extremely subtle. In the space we have allowed ourselves it is difficult to give more than the crudest sketch of the approach.

Science is concerned with an understanding of the world and how it works. This understanding has traditionally developed through open discussions and controversies among scientists. Bohr (and incidentally the twentieth-century philosopher Ludwig Wittgenstein) observed that all our discussions take place within 'everyday' language and that language is the only means of communication for the fulness of our thoughts.

Language has evolved through man's interactions with society and the natural world – animals, chairs, and so on. It is inseparably bound up with the world we live in, which is filled with objects described by 'classical' or Newtonian physics.

It is not surprising that we experience no difficulty in matching a verbal description of a billiards match with its scientific description in mathematical terms. When we come to the world of atoms and elementary particles, however, new laws of physics are needed and classical mechanics is replaced by quantum mechanics. It is at this point that Bohr cautioned us with what has become known as the Copenhagen interpretation of quantum mechanics. He pointed out the pitfalls that lie in wait for anyone who attempts to press day-to-day language, which is adapted perfectly to our large-scale world, into the world of quantum particles. Bohr suggested that physicists should renounce the use of 'pictorial descriptions' and remarked on the paradoxes and confusions which would arise if they attempted to talk about atoms as miniature billiard balls. 'Does the electron really have a path when we are observing it?' is the sort of question which is doomed to end in confused argument.

Bohr pointed out, however, that experiments on atoms involve large-scale apparatus, such as accelerators, Geiger counters, cloud chambers, and electromagnets. The results of an experiment, which can be expressed as the click of a Geiger counter or the movement of a needle on a dial, are entirely classical. Provided scientists confine themselves to such observations there is no difficulty in discussing atomic experiments in everyday language. In this way a successful mathematical theory, quantum mechanics, correlates the results of various experiments and this can be discussed unambiguously in ordinary language. It is only when scientists go beyond this description and attempt to draw a detailed 'picture' of an elementary process or particle that confusion arises.

The Copenhagen interpretation therefore tells us how science should deal with the world of atoms without becoming embroiled in paradox and warns of the consequences of pressing language into an area for which it was not adapted. In this respect Bohr's interpretation echoes the writings of Wittgenstein, who proposed that many of the 'great questions' of philosophy were simply confusions generated by an inattentive use of language. Wittgenstein believed that the problems of philosophy had their origins within the misuse of language rather than in some metaphysical realm. It is the business of both science and philosophy to proceed with care and avoid pseudo-problems which are brought about by pressing language beyond its limitations.

COMPLEMENTARITY

The notion of complementarity was proposed by Niels Bohr, who felt that its significance extended far beyond the intellectual confines of theoretical physics. In essence complementarity states that an object or process cannot be pinned down within a single description; rather several overlapping and possibly incompatible descriptions are needed to exhaust its variety. Two crude analogies for complementarity can be given: several photographs taken from different locations and angles which give an overall description of the exterior of a building, and the varying accounts from wife, friends, and children that piece together a man's life.

Bohr felt that the ideas expressed by complementarity were closer to the working of nature and our minds than those dictated by 'common sense' or 'classical logic.' When two descriptions of an event do not overlap common sense would dictate that an error has been made; Bohr's complementarity, however, would indicate that a profound truth is being approached.

NON-CLASSICAL LOGIC

A scientific theory can be analysed to determine the logical structure of its arguments. If the theory can be written down in such a way that it only asks questions which can be answered by a simple 'yes' or 'no,' then it is based on classical logic. If this cannot be done, then the theory has a 'non-classical logic' at its foundation. For example, a theory which involves complementary descriptions has moved beyond classical logic.

For some thinkers the introduction of non-classical logics into physics represented a step away from a rational response to the world. It was felt that classical logic has a special position in thought, and its abandonment introduces an unnecessary 'mystical quality' to science.

THE COLLAPSE OF THE WAVE FUNCTION

The so-called collapse of the wave function is a problem which confuses many students of quantum theory. Not only students but eminent physicists become puzzled by it, for it has led to such paradoxes as 'Schrödinger's cat,' 'Wigner's friend,' and dialogues·between Bohr and Einstein on the completeness of the quantum mechanical description of nature.

A scientific account of processes within the large-scale world is obtained through Newton's equations of motion. The mathematical solutions to these equations give information on the position and velocities of objects such as billiard balls and stones. To know the exact way in which billiard balls move during a game is to have a full description of that game, and this is inherent in Newton's equations of motion.

In the world of atoms things are not quite as simple. The solution to the equations of quantum mechanics is called the wave function. Rather than giving an exact position and velocity for a particle, the wave function provides the probability of finding the particle within a given location. The quantum mechanical account of an experiment on an electron, for example, may be given as a series of probabilities of finding the electron at different locations in the laboratory. These probabilities are not a measure of ignorance or experimental error but are a fundamental expression of the nature of atomic matter.

Einstein objected to this aspect of quantum theory by using the following argument. Suppose that during an experiment an electron enters a Geiger counter and causes a 'click' to be heard. Then at that instant we are 100 per cent certain that the electron is not located anywhere else but within the Geiger counter. However, a moment before the Geiger counter clicked the mathematical solutions that followed from quantum theory showed that the wave function was delocalized over the whole laboratory. Hence at the instant of the click the wave function must have 'collapsed' inside the Geiger counter. Einstein argued that sudden change is not described in the mathematics of quantum mechanics and that the theory must be incomplete.

Niels Bohr replied, within the spirit of the Copenhagen interpretation, that we must be careful to discuss only the objective parts of the theory which are observable results of experiments. The wave function, however, is a mathematical device which has no objective or measurable existence. The discontinuous change in the wave function upon the electron registering its appearance by a click in a Geiger counter is simply a change in the description of the experimental situation. After the Geiger counter has clicked the system has obviously changed.

For some scientists and philosophers the collapse of the wave function still remains a stumbling-block to the acceptance of the Copenhagen interpretation whereas for others it is simply a pseudo-problem.

Glossary

Bateson, Gregory An original thinker who has made contributions in various fields, including anthropology, psychiatry, and cybernetics. His writings have stressed the preoccupation of the mind with perceiving differences and differences of differences. This notion led Bateson to his Double Bind theory of schizophrenia, which was later applied by R.D. Laing in his studies of family influence in mental illness.

combinatorics A branch of mathematics concerned with the packing and arranging of patterns and designs and with combinations and permutations. Roger Penrose used combinatorics in the study of large networks of spinors in an attempt to derive the properties of space.

commutation In mathematics, the interchange of order of two quantities added, multiplied, etc. For the natural numbers the order of multiplication does not affect the result. The order is significant, however, when matrices (q.v.) are multiplied together. Matrices whose products depend on the order of multiplication are said to be non-commuting. The results of two quantum mechanical measurements generally depend on the order in which they are carried out; that is, they are non-commuting.

complementarity An idea propounded by Niels Bohr that nature is so rich that a single description will be insufficient to exhaust the definition of a phenomenon.

conformal invariance A property attributed to an equation or theory which is unchanged by the operations of the conformal symmetry group. The conformal group is an extension of the Lorentz group (q.v.) and contains all symmetry operations in space-time which leave the light cone (q.v.) unchanged. The group relates to massless particles in special relativity.

continuum Loosely speaking a continuum of numbers occurs when between any two numbers, no matter how close they are chosen, there can be found an infinity of other numbers. The natural numbers form a continuum, but the integers do not.

dissipative structures Statistical mechanics (q.v.) is often taught as that branch of science in which chance reigns supreme and structures are doomed to erosion through random fluctuations of their constituents. In contrast Ilya Prigogine points out that nature throws up stable complex structures which are capable of adaptation and survival. He sees such *dissipative structures* as arising in 'open systems' through the free exchange of energy and materials with the environment.

Einstein-Rosen-Podolski paradox Several of the founders of quantum mechanics had misgivings about the theory and the Copenhagen interpretation (see Appendix). To make their doubts more concrete Einstein, Schrödinger, Wigner, and others devised hypothetical situations (*gedanken* experiments) which lead to paradoxes when discussed. Bohr denied that such paradoxes existed and believed that such *gedanken* experiments, of which the Einstein-Rosen-Podolski experiment is one example, were capable of unambiguous interpretation.

entropy A quantity occurring in thermodynamics and statistical mechanics which relates to the 'disorder' present in a system. Unlike temperature, pressure, energy, and mass the entropy of a system cannot be measured directly but is inferred from other quantities.

genotype, phenotype The *genotype* is the total genetic information possessed by an organism. As the organism develops and interacts with its environment so a certain amount of its genetic potential (genotype) finds expression as size, shape, colour, behaviour, etc. This individual manifestation of the genotype by a particular organism or group of organisms is called the *phenotype*.

Gödel's theorem A consequence of the study of the foundations of mathematics and other deductive systems by the mathematician Kurt Gödel, the theorem states that certain deductive systems are inherently incomplete, in the sense that there always exist propositions within the system which cannot be proved. The implications which have followed from this theorem are often imaginative but controversial. It has been suggested, for example, that the theorem implies that the ability of computers will never duplicate that of the human brain.

Hamilton-Jacobi theory Classical mechanics is often presented as being based on Newton's laws of motion but the same results can be obtained using different assumptions and starting-points. One of these is the Hamilton-Jacobi theory, whose equations appear far more abstract than those of Newton's mechanics. Discussions of classical mechanics in the

Hamilton-Jacobi form illuminate correspondences with quantum mechanics.

invariance A property attributed to the equations of a particular theory (Maxwell's equations, Schrödinger's equation, field equations of general relativity, etc.) which are unchanged by a variety of symmetry transformations. The equations are then said to be invariant with respect to the operations of that particular symmetry group (q.v.).

isospin In the early days of quantum theory it was discovered that the electron possesses a two-valued degree of freedom, its spin (q.v.), in addition to its other degrees of freedom. It was later found that the idea of spin symmetry in space could be extended to include a degree of freedom corresponding to spin in an abstract space (isospace). By introducing the notion of isospin it became possible to consider two different particles as a single particle possessing different isospin states. The relationship between an abstract or 'internal' symmetry such as isospin and the symmetries of space-time is not clear.

least action One of the possible formulations of classical mechanics (see Hamilton-Jacobi theory). Whereas Newton's equations of motion build up the movement of particles by considering the instantaneous forces present at each element of their path, the principle of least action is based on an over-all property of the motion – that the particle assumes a trajectory that will minimize its 'action.'

light cone The volume traced out in space-time from a source of light, which might be thought of as the 'history,' in space-time, of a light beam. Two space-time points which lie within each other's light cone are causally connected because they can exchange signals and experience each other's influences. Two points which lie outside each other's light cone cannot influence each other in any way since signals moving faster than the velocity of light would be required to connect them.

Lorentz group The group of symmetry operations in space-time which leave the laws of nature unchanged in the special theory of relativity.

matrix A mathematical object consisting of an array of numbers. Matrices obey different rules of operation than do ordinary numbers and, when multiplied, do not generally commute (q.v.). In Heisenberg's formulation of quantum mechanics the operations or experiments of quantum theory are represented by operations on matrices and the experimental results by the numbers found in these matrices.

propositional calculus The laws of logic could be thought of as a set of procedural rules together with initial assumptions. If each proposition is represented by a mathematical symbol and the rules of procedure by mathematical operations then a logical discourse can be represented by a set of symbolic manipulations. This symbolic form is called the propositional calculus. The propositional calculus has been used to analyse the statements of quantum theory.

Riemannian geometry Bernard Riemann investigated geometries which are more general than those assumed in Euclid. The methods of Riemannian geometry were used by Einstein in the mathematical formulation of the general theory of relativity in which space-time possesses curvature and is non-Euclidean.

Russell's paradox Formulated during Bertrand Russell's investigations on the foundations of mathematics, the paradox concerns self-referential systems and can be stated as follows: if R is the set of all sets that do not belong to themselves, does R belong to itself? An informal statement of the paradox is made in terms of a barber in a certain village: This barber is the man who shaves all men who do not shave themselves. Who shaves the barber? Biological systems have sometimes been discussed in these terms; for example, DNA contains all the genetic material that describes a system, and included in this description is a description of DNA and all the information it contains.

spinors, twistors Spinors are mathematical objects that correspond to the electron's two-valued spin in quantum theory; spinors are also used in relativity theory. The twistor is a mathematical generalization of the two-component spinor made by Roger Penrose; it possesses four components and is of value in exploring the connections between quantum and relativity theories.

statistical mechanics Solids, liquids, and gases appear very different from the atoms and molecules which compose them. When attempts were first made to derive the properties of macroscopic systems from their constituents the astronomical number of entities involved made 'exact' calculations impossible. It was therefore decided to treat the motions of atomic particles in a statistical fashion using statistical mechanics and derive macroscopic properties such as pressure and temperature through averaging processes.

superposition principle A principle applied in quantum mechanics, where any linear superposition of allowable states of a system is itself an allowable state and, conversely, any state contains components from all other states.

symmetry breaking An idea which has become fashionable in modern physics, symmetry breaking occurs when a stable (or ground) state of a system appears to violate the symmetries which are present in the physical equations which govern it. For example, the equations which govern magnetic matter are spherically symmetric yet a ferromagnet violates this symmetry by exhibiting a preferred direction in space – the direction of its magnetic axis. Attempts have been made in particle physics to relate symmetry breaking to the appearance of certain particles. One of us (DP) has attempted to relate symmetry breaking in large quantum systems to the manifestation of classical variables.

symmetry group A mathematical group containing the various symmetry operations (rotation through a certain angle, reflection about a certain axis, translation over a certain distance) which leave the appearance of an object or an equation unchanged.

Wittgenstein's theory of language Ludwig Wittgenstein, an Austrian philosopher who spent much of his creative life at Cambridge, was preoccupied with language and in the *Tractatus Logico-Philosophicus* attempted to fix the boundaries of unambiguous philosophical argument through a theory of language. Language was believed to 'picture' the world, and the domain of philosophy was the analysis of scientific propositions within this picture. In his later life Wittgenstein pointed out the limitations of his 'picture' theory of language and stressed the richness and variety of language and spoke of 'language games.' In *Philosophical Investigations* he concludes that many of the traditional problems of philosophy have arisen because language has been used in an insensitive fashion.